THE SEVEN DEADLY MISCONCEPTIONS

THE SEVEN DEADLY MISCONCEPTIONS

H V Mohanlal

Notion Press

Old No. 38, New No. 6
McNichols Road, Chetpet
Chennai - 600 031

First Published by Notion Press 2016
Copyright © H V Mohanlal 2016
All Rights Reserved.

ISBN 978-93-5206-911-8

This book has been published with all efforts taken to make the material error-free after the consent of the author. However, the author and the publisher do not assume and hereby disclaim any liability to any party for any loss, damage, or disruption caused by errors or omissions, whether such errors or omissions result from negligence, accident, or any other cause.

No part of this book may be used, reproduced in any manner whatsoever without written permission from the author, except in the case of brief quotations embodied in critical articles and reviews.

Dedication

This book is dedicated to my late parents and all the scientists, physicists, philosophers, professors, intellectuals, and others from the past to the present for working tirelessly to understand nature and unravel its secrets not only for the betterment of the society as a whole but also to ultimately unravel the secrets behind the mystery of life and the purpose of our existence.

Contents

Acknowledgment	*ix*
Student's queries for which physics has no cogent answers	*xi*
Preface	*xix*
My Observations	*xxi*

CHAPTER ONE	1
1. Introduction to false facts (misconceptions)	3
2. The moon paradox	6
3. Discussion on Einstein's lecture on the topic of ether and relativity	21
4. ETHER	25
5. Quotes on ether	27

CHAPTER TWO	33
1. Misconception number 1	35
2. Misconception Number 2	59
3. Misconception number 3	61
4. Misconception number 4	65
5. Misconception number 5	91
6. Misconception number 6	163
7. Misconception number 7	169

CHAPTER THREE — 173

1. Forces that hold together bodies right from atoms to black holes — 175
2. Theory of everything — 177
3. Inferences — 182
4. Recapitulating the list of already existing evidence and the proposed experiments — 184
5. Speed of light — 186

Epilogue — *189*

Glossary — *192*

References — *193*

Acknowledgment

I would first like to thank Discovery Science channel, as the program aired in this channel inspired me to write this book. I also thank the people responsible for inventing the wonderful thing called the Internet. My special thanks to Wikipedia and the people who uploaded the vast amount of information without which I could never have written this book. I also thank all other people and my publisher for assisting me in this endeavor. I also thank my wife Sunanda for her constant support, encouragement, and firm belief in me.

Student's queries for which physics has no cogent answers

> *"The most important thing I found out (from my father) is that if you asked any question and pursued it deeply enough, then at the end there was a glorious discovery of a general and beautiful kind."*
>
> Richard Feynman

The following few passages are from the article titled *Science Is At Its End* written by William J Beaty, an American researcher engineer:

It seems that every so often, a fairly large group of scientists begin to assert that science is just about complete, that the vast unknown is gone, and that all the major research can stop because we now known everything except the details.

Example quotes

> *"There is nothing new to be discovered in physics now. All that remains is more and more precise measurements."*
>
> Kelvin Lord William Thomson (1900)

> *"The more important fundamental laws and facts of physical science have all been discovered, and they are now so firmly established that the possibility of their ever been supplanted in consequence of new discoveries*

is exceedingly remote. Our future discoveries must be looked for in the sixth place of decimals."

Albert A. Michelson of the famous Michelson—Morley experiment (1894)

Beaty further states that it is the conceit of every age to believe that scientific advancement has at last reached its pinnacle, while future explorers will have very little to do. So, in order to open the way to revolutionary discovery, you must reject this conceit.

Alternate views

"I think that it is a kind of intellectual chauvinism to assume that all the laws of physics have been discovered by the year of our meeting. Had we held this meeting twenty or forty years ago, we would perhaps have erroneously drawn the same conclusion."

Carl Sagan (1971)

"I have begun to feel that there is a tendency in the 20th century science to forget that there will be a 21st century and indeed a 30th century from which vantage points our knowledge of the universe may appear quite different than it does to us. We suffer, perhaps, from temporal provincialisms, a form of arrogance that has always irritated posterity."

J. Allen Hynek (1966)

It is indeed naive of anyone to think that they have discovered everything in science, especially physics. The following questions, which lack cogent answers, are a clear testimony to this fact.

Questions on atoms

1) Nature has created everything existing in this universe with a specific purpose and task. Hence, what is the purpose of the empty space present in atoms?

2) Is the empty space in an atom the same as that existing in outer space?

3) When atoms of lighter elements, which are in a free state, fuse together to form atoms of heavier elements, their volume does not increase proportionately. What happens to the empty space associated with the atoms of the lighter elements after they fuse together?

4) What happens to the empty space associated with a massive star that transforms into a neutron star or a stellar black hole measuring only 10–15 km in diameter after a supernova has occurred?

5) If an atom consists of mostly empty space, what prevents objects from passing through each other?

6) From when and on what basis exactly did physicists start believing that ordinary matter is a condensate of electromagnetic energy only and if this assumption is true then, why can't electromagnetic energy be converted into ordinary matter?

7) If atoms consist of only protons, neutrons, electrons, and 99.99% of empty space, why do objects appear as being full and what is the low-density substance present in all objects? Is the FBBTS (flesh, blood, bones, tissue, and skin) present in our bodies just an illusion created by the whirring electrons or electron clouds?

8) Each addition of a proton to an element results in the formation of the next element with a higher atomic number. What is the internal mechanism or process by which this transformation takes place?

9) When an object like wood burns, what exactly is burning? Is it the protons, neutrons, or electrons?

10) Since, as per current theories, electrons, which are negatively charged and form the outer shells of atoms can exist either as point particles spinning rapidly around the nuclei or as electron clouds how do atoms manage to stick to each other?

11) Why is the range of the strong nuclear force, which is billions of times stronger than gravity, so unimaginably and ridiculously short? It is effective only within the nucleus of an atom.

12) What type of a force is the strong nuclear force electromagnetic, electrostatic or mechanical?

Questions on gravity and outer space

1) Why is matter, by itself, considered weightless?

2) How exactly does mass gives rise to the force of gravity or the gravitational field?

3) Under which category is the gravitational force classified, mechanical or electromagnetic?

4) Mass is defined as the amount of matter contained in an object. It is also defined as the resistance an object offers to being accelerated when it is subject to a force. Even in a gravity-free environment, we have to apply a force to overcome inertia and move the object. What characteristic or property of matter gives rise to this resistance or inertia?

5) Think of a rocket whose mass is say 100 tons and weight 980 KN. As its distance from the earth's surface increases, its weight, according to current theories, will gradually decrease as the strength of gravity is also decreasing, and at a particular altitude, say 1.5 million km, its weight will be almost zero although mass remains unchanged at 100 tons. Hence, theoretically, a thrust of only 3–4 N or even less is sufficient to propel the rocket upwards. However, as the rocket is still in possession of its mass, is it not necessary to apply an appropriate force to overcome its inertia in the upward direction also or is inertia applicable only in the horizontal direction?

6) Tremendous amount of energy is required to accelerate any object. Because gravity is supposed to be a force, which causes freely falling objects to accelerate, what is the source of this energy, and if such a force really exists, what is the reason behind its non-detection?

7) Are gravity waves, hypothesized to exist by Einstein, electromagnetic or mechanical in nature?

8) How can the range of gravity be infinity if it is the weakest of the four fundamental forces?

9) Why are most celestial bodies spherical in shape, and irrespective of their mass, why do liquid bodies also attain a spherical shape in outer space?

Student's queries for which physics has no cogent answers

10) Why liquid bodies do not assume a spherical shape below 100-km altitude from the earth's surface?

11) Why does weightlessness or microgravity occur only above 100-km altitude from the earth's surface?

12) 99.99997% of atmospheric air exists below 100-km altitude from the earths' surface and the remaining 0.00003% spreads out in thousands of kilometers above this altitude implying that outer space is just an empty vacuum. If this is true, why do spacecrafts encounter severe orbit decay, which is highest just above 100-km altitude and decreases with increase in altitude?

13) During re-entry, why do spacecrafts heat up exactly at 120-km altitude from the earth's surface where the density of air is almost negligible?

14) Similarly, during re-entry why do spacecrafts, like the shuttle weighing hundreds of tons, bounce back into space at exactly 120-km altitude from the earth's surface if their angle of attack (AOA) is less than 40°?

15) Why do objects and astronauts inside a sounding rocket, which does not go into a regular orbit, experience weightlessness in space even while ascending?

16) The strength of gravity on the moon's surface is one-sixth the value of earth's gravity. As numerous manned and unmanned missions have landed on the moon, did the astronauts conduct any direct weighing experiment to verify this postulate? If yes, what was the result and if not, why did they not conduct it?

17) When a student questions how gravity can bend light if the rest mass of photons is zero (gravity acts only on bodies or particles that possess mass) he gets the following reply. Since energy and mass are equivalent, gravity is able to bend light. However, since this explanation violates the laws of quantum mechanics, according to which gravitational interaction at the quantum level is nonexistent, how does one explain this dichotomy? The photons are so unimaginably small that the total number which travel from the sun and strike us every second are actually spread across a length of 300,000 km, which is light's speed. Moreover, if this assumption that the photons are zero mass particles is true, where is the suns' mass disappearing and how are they able to gain momentum to travel at the speed of light?

Student's queries for which physics has no cogent answers

18) Because there are no upward and downward directions in deep space, will the astronauts inside a spacecraft traveling outside the gravitational field of the sun feel that they are traveling in a horizontal or in a vertical direction.

19) As absolute empty space, which is a vacuum, and time are not physical entities in themselves, how exactly do they merge and form the mysterious entity called space-time?

20) Is space-time a physical entity, and if yes, how does mass affects its curvature? If it is not a physical entity, then what exactly is it?

21) Why are most contemporary scientists dead scared of ether theories? After all, if it is there, it is there, and if not, no. From when and why are these theories banned?

Questions pertaining to our solar system and general questions

1) Why are huge celestial bodies always gaseous in nature and smaller heavenly bodies like earth always rocky?

2) Why does the distance between planets in our solar system steadily increase, almost doubling starting from Mercury until Neptune?

3) What natural forces initially launched all celestial bodies into their orbits and made them rotate?

4) Why do scientists believe that antimatter is missing from our universe?

5) Earth, a ridiculously tiny celestial body, contains all the elements present in the periodic table of elements, whereas the trillions of stars present in the billions of galaxies are supposed to contain only hydrogen and helium along with a few traces of lighter elements. What could be the reason for this strange anomaly?

6) Distant planets like Jupiter, Saturn, Uranus, and Neptune are releasing more thermal energy than they receive from the sun. What is the source of this excess energy?

7) According to the theory of the birth and evolution of our universe, a number of nonmetallic stars (in astronomy, other than hydrogen and helium, all other elements are metallic) should have been present in

the universe. Hence, why have astrophysicists been unable to discover a single nonmetallic star even until today?

8) Why do solar systems and galaxies exist only in a flat plane?

9) Do black holes really have a hole in them? If yes, how does this hole come into being?

10) Why do galaxies mostly appear to be spiral shaped?

11) Does time really exists or is it just a notion?

12) According to the current theories, the speed of light is always constant and independent of the motion of its source. To satisfy this condition, the speed of light emerging from the moving source should increase when it is traveling opposite to the direction of motion and decrease when it is traveling in the same direction. Is this presumption true? If it is not, then how exactly does light regulate its own speed?

13) Light by itself is invisible. However, when it strikes an object, not only does it become visible, it also generates heat. What exactly happens when it strikes an object?

Conclusion

Although these are pertinent questions, they are conveniently ignored, and hence, when students ask such questions, they receive vague answers or are sometimes gently rebuked hinting that the questions are irrelevant and silly. However, far from being silly, these questions are very important and relevant because the answers to these questions may help resolve the various problems and mysteries existing in the world of fundamental physics. These questions also make it clear that something vital is missing from the current theories and that is with respect to matter. With the current theories, even the most brilliant of persons cannot explain the manner in which electromagnetic energy transforms into the low-density matter contained in all objects including our bodies.

Focusing now on to new discoveries as long as one conforms, that is, works within the boundaries of current theories, the maximum one can dream to achieve is to win a Nobel Prize. However, if your aim is to achieve something greater, then you have to not only ask similar questions as mentioned herein but also have to find proper answers for them. For that, you will have to abandon the beaten track and tread new and unchartered territories. This may endanger not only your career but also alienate your colleagues and friends. However, it is also possible that this new path will lead you to a pot of gold and make everyone who doubted you to eat humble pie as has, often happened in the past.

Preface

"Your time is limited so don't waste it living someone else's life. Don't be trapped by dogma—which is living by results of other peoples thinking. Don't let the noise of others opinions drown out your own voice, and most important, have the courage to follow your heart and intuition. They somehow already know what you truly want to become. Everything else is secondary."

Steve Jobs

In the early months of 2012, I was watching a popular TV program on the Discovery Science channel, *Through the Wormhole with Morgan Freeman*. In the course of that program, a detailed explanation followed as to how space-time curves around huge heavenly bodies. Next, an example, in which a ball representing the sun created a depression when placed on a trampoline, illustrated the concept of curved space-time. The man behind this theory, known as the general theory of relativity, was the famous scientist, Albert Einstein.

After the program, I thought about this concept of space-time where time represents the fourth dimension. I was completely baffled by this because until then I was under the impression that outer space was just an empty vacuum and time is a notion. Because both of them were not physical entities, in the normal sense, I wondered how they could interact with each other and curve the vacuum of space itself.

As this problem kept bugging me, initially I started researching the related topics on the Internet out of sheer curiosity. However, over time, it became a sort of an obsession. I spent more hours on the Internet searching for answers and solutions for the numerous questions that kept popping up in my mind. I went about it, neglecting even my business, as if someone had entrusted me with an important job and it was my duty to complete it. In the course of my research, I gained a vast knowledge about the old and new theories regarding gravity and other mysteries of the universe.

After reading numerous articles and quotes written by renowned personalities like Rene Descartes, Charles Darwin, Galileo, Sir Isaac Newton Newton, Euler, Maxwell, Einstein, and a host of others, I began to doubt, question, and seek the truth about everything that sounded illogical and is contrary to what we perceive and experience in our daily lives. Prior to this, I firmly believed that science is all about absolute truth, logic, and facts that are established after thorough research, experimentation, and verification. I presume that everyone else also must be holding the same perception because even the most brilliant of scholars, professors, and scientists of today including all others have passed through the student stage. From our childhood, we unquestioningly believe in everything we are taught, even though certain things may sound illogical, and as we grow older, this belief is further strengthened.

My Observations

> "*Have no respect whatever for authority; forget who said it and instead look at what he starts with, where he ends up, and ask yourself "is it reasonable?"*
>
> Richard Feynman

1) All that we study in educational institutions, which is taught as if it is the sacrosanct truth, is not necessarily 100% true because theories, laws, and so on have been written taking into account various information and knowledge available at that point in time and from the perception of the individuals concerned. Hence, with the passage of time, all theories are subject to modification as and when new evidence and knowledge are gained. Hence, it is not necessary that the majority view or the official view is always correct.

Examples

a) For thousands of years, it was believed that the earth was at the center of the universe. Now we know that this is not true.

b) Atoms that make up matter were thought to be full of a gel-like substance. However, it is now an established fact that they contain an extremely dense and unimaginably small nucleus, which is supposedly surrounded by empty space and electrons.

c) The universe was thought to be slowing down because of the pull of gravity. However, on the contrary, the Hubble telescope has discovered that the universe is actually expanding rapidly.

d) The universal laws of gravitation, which states that every object attracts every other object in the universe, have been taught to all of us as if it is the absolute truth. However, after Einstein's publication of general relativity, the concept of gravity has changed dramatically. According to Einstein's theory, which is supposed to have superseded Newton's, gravity is not a true or independent force but a curvature of space-time. However, most of us even now, are comfortable with Newton's than Einstein's theories because they are simple and easy to understand.

2) After I learnt that, the gravitational force had not yet been discovered I began to ponder about it and observed that it was not only a mysterious but also a magical and intelligent force. Although it is the weakest force, its range is infinity, it cannot be shielded against, it is always attractive, and it is able to distinguish the different physical features of objects thus enabling it to act only according to mass and not surface area. Moreover, if such a force really existed, an unimaginable amount of energy is required to give rise to and sustain it, which however, has not yet been discovered. That is why scientists have gradually abandoned the spooky action at a distance theory and adopted the gravitational field theory. However, even here, the exact manner in which mass gives rise to the gravitational field is a mystery.

The belief that gravity alone endows weight to matter and forces freely falling objects to accelerate is so deeply entrenched in our minds from generations that we do not, even for a second, think otherwise. What if weight was is an inherent property of matter? (After all other than endowing weight to matter gravity has no other function.). Even then, objects on the earth's surface would fall down because of their heaviness and accelerate due to the potential energy stored in them. In such a case, the mystery behind gravity and the various magical properties associated with it would vanish instantly. We would only need to find out why the rate of acceleration during free fall varies with altitude. Excited by this thought I went to great lengths to find out why matter, by itself, is considered weightless, but strangely, I could not get a logical answer because nobody really knows why. However, I ultimately managed to figure it out.

3) According to Einstein's equivalence principle, concerning acceleration and gravity, an astronaut standing inside a spacecraft's cabin, which is completely closed and accelerating upwards at the rate of 9.8 m/sec^2 in a gravity-free environment, will experience a gravitational force of one G, which is equal to that experienced on the earth. Hence, it

would be impossible for him to tell whether he is standing inside the cabin of a spacecraft or on the earth's surface. Einstein claimed that, *"no experiment, no clever exploitation of the laws of physics can tell us whether we are in free space or a gravitational field."* As I thought about this claim, I suddenly realized that by simply jumping up the astronaut could instantly find out his location. When he jumps on the earth's surface, he will feel himself going up in the air and then coming down to touch the stationary ground. However, when he tries to jump up when he is inside a spacecraft accelerating upwards at 9.8 m/sec^2 his feet will never even leave the spacecrafts floor because the velocity of the floor along with the spacecraft is constantly increasing. To rise above the spacecrafts floor he should jump up with an acceleration, which is greater than 9.81 m/sec^2.

Einstein came up with the equivalence principle after he realized that during free fall, in a uniform gravitational field, a person would be unable to feel his own weight as she/he is falling along with the gravitational field. Initially, the importance of this thought experiment escaped my attention, but later on, I was elated when I realized that this same analogy could be used to perform a simple practical experiment by which we can decisively find out if an independent gravitational force existed or not. If only Einstein had thought about this, he could have proved, that his hunch, that an independent gravitational force did not exist was indeed true at that point in time itself. I also thought of a few other experiments, which could positively prove that an independent gravitational force does not exist. The chapter on gravity and the following link on YouTube *The mystery of gravity by Mohanlal* contain a detailed description of the experiments.

4) Just as the long range of gravity defies logic, the extremely short range of the strong nuclear force, which is supposed to be billions of times stronger than gravity, is also inexplicable. Its range is ridiculously short; it never extends beyond the nucleus of an atom. Logically speaking, all forces must behave and function similarly because nature functions in a highly methodical, logical, and repetitive manner rather than in an ad hoc or whimsical way. This means that the same force that holds an atom's nucleus together must also be holding celestial bodies as they are nothing but a huge conglomeration of atoms. That is why both gravity

and the strong nuclear force have one important feature in common, which is that, unlike the electromagnetic forces, both act on all matter in general strongly implying that both arise from the same cause. Another strange feature about the atom is that although it is supposed to be comprised of almost 99.9% empty space, ordinary matter appears anything but empty. When most metals like, for example, iron are cut and drilled, they do not appear to be empty but full of matter. If matter was truly, mostly empty space, it should have appeared as a rigid sponge and not as solid matter.

5) There are many unexplained phenomena, which occur in nature, and even on our own planet earth, which if investigated properly may reveal valuable information. However strangely, these phenomena do not receive the attention they deserve. We should not forget that although the earth along with its atmosphere is constantly evolving with the passage of time, outer space, which begins from above 100-km altitude from the earth's surface, remains unchanged and is probably as old as the universe itself.

Examples

a) When a spacecraft weighing hundreds of tons (e.g., a space shuttle) is re-entering earth's atmosphere, it bounces back into space at exactly 120-km altitude from the earth's surface if its AOA is less than 40° to the vertical plane. This phenomenon, which is exactly similar to how a flat stone thrown at a low angle bounces on a placid lake's surface, is a real mystery and has no logical explanation. This is because 99.9997% mass of earth's atmosphere exists below 100-km altitude and the remaining 0.0003% is spread out in the thousands of kilometers that make up the thermosphere and the exosphere. Even if we suppose that there is sufficient air at that altitude to cause friction, then heating should have commenced gradually and then steadily increased as the spacecraft descended deeper and deeper into the denser atmospheric air instead of bouncing back into space like a rubber ball. Moreover, it is a well-established fact that the actual heating due to air occurs below an altitude of 80 km from the earth's surface. Strangely, the aircraft does not heat up between 80 and 100-km altitude.

b) Orbiting spacecrafts are considered to be constantly falling around the earth's surface, like a freely falling elevator, and hence weightlessness experienced by astronauts and objects while inside these spacecrafts is attributed to it. This assumption is highly questionable because weightlessness occurs even inside a sounding rocket, which flies vertically upwards into space and then drops down. Moreover, although an elevator falls at 90° to the earth's surface objects within, it will not experience weightlessness even though the scale registers their weight as zero and neither will liquids take on a spherical shape as they do in space. In contrast, an orbiting spacecrafts trajectory at all times is parallel to the earth's surface. From these observations, it is clear that the real cause behind weightlessness in space, also referred to as microgravity, is still a mystery.

6) Once a scientific theory is accepted and approved by peer review, and is taught in educational institutions, it becomes impossible, even for the person who originally wrote or discovered the theory to change or modify it though it may be of vital importance. Because of this, the establishment rejects their later views and theories due to which they are not included in the curriculum of educational institutions thereby depriving future generations of this information, which may be of vital importance.

Examples

Sir Isaac Newton wrote the universal laws of gravitation, according to which gravity is an independent attractive force, when he was about 45 years old. However, as he was unable to explain the mechanics behind the spooky action at a distance, an older and wiser Newton published a few alternate theories on gravity. Prior to that, he had also written a letter to Robert Boyle at the end of the seventh century, which Bibliotheque Raisonnee published in 1745 in which Newton categorically denied that bodies attract one another across empty space. He also suggested that the real cause behind this phenomenon was an all-pervasive mechanical ether and also postulated that the new ether consisted of particles endowed with very powerful short-range repulsive forces. The ether had to be extremely rare so that it would not obstruct the motion of planets and yet very elastic and springy so that it could push large masses toward one another. In his later theories, he also posited that the density of ether near celestial bodies is higher and decreases as the distance from the body increased, which would give rise to refraction causing starlight passing closer to massive celestial bodies to bend. In this case, Newton treats gravity as a field and not as an instant spooky action at a distance. However, as he had no proof to substantiate his claims, the new theory failed to gain acceptance.

b) Einstein's theory of special relativity dealt a severe blow to the various ether theories, which gradually lost relevance and is now taboo to most modern scientists. However, he again reintroduced it in his 1920 lecture in the University of Leiden, Holland, in which he even went to the extent of stating that space without ether is unimaginable because then it would possess no physical qualities whatever. But he failed in his valiant attempt of resurrecting the ether theories as the then contemporary scientific community, which otherwise not only held him in high esteem but also treated everything he said or posited as the gospel truth, strangely rejected this particular claim.

We will see, later on in this book, how the rejection of these two claims has seriously hampered the progress of fundamental physics.

7) Philosophy has lost its relevance in physics, or for that matter any other branch of science, as it has become highly specialized, mathematical, and technical due to the use of sophisticated instruments, supercomputers and so on. The following words adapted from the book *A Brief History of Time* by Stephen Hawking aptly describes the present situation, *"Up to now, most scientists have been too occupied with the development of new theories that describe what the Universe is than to ask the question why. On the other hand, the people whose business it is to ask why, the philosophers, have not been able to keep up with the advance of scientific theories. In the eighteenth century, philosophers considered the whole of human knowledge, including science, to be their field and discussed questions such as, "Did the Universe have a beginning?" However, in the nineteenth and twentieth centuries, science became too technical and mathematical for philosophers, or anyone else except a few specialists. Philosophers reduced the scope of their enquiries so much that*

> *Wittgenstein, the most famous philosopher of this century, said "The sole remaining task of philosophy is the analysis of language." What a great comedown from the great tradition of philosophy from Aristotle to Kant!"*

It is my humble opinion that as far as fundamental physics is concerned, the physical reality of our existence, sound logic, and the philosophical views of our illustrious ancestors should be taken into consideration along with the latest scientific discoveries being made today, wherein new theories are proposed. Failing to do so is like trying to paint a masterpiece without a brush. We will always get an imperfect or untrue picture of things. Philosophy, which originates entirely from the human mind and which in turn is tuned in with mother nature in a mysterious and harmonious manner, bridges the gap between hard (modern) science, and physical reality as modern science especially physics is almost entirely dependent on instruments and computers.

Examples of the importance of philosophy

a) Demetrious, a Greek philosopher, predicted the existence of the atom in 400 BC, which has now turned out to be absolutely true. Dalton, father of the atomic theory, and many others must have definitely been inspired by this philosophical thought.

b) Ancient Indian and Chinese philosophical texts, written thousands of years ago, predict that the universe is a product of two different entities or opposite energies. The Greek philosopher Heraclitus also propounded the theory of dualism or opposites. Today we know for sure that matter contained in the nucleus of an atom is a condensate of electromagnetic energy and thanks to Paul Dirac, the opposite of matter that is antimatter has also been discovered. However, why it seems to be missing from our universe is one of the greatest mysteries in physics.

The important question here is how people living thousands of years back could have guessed about the atom or the second matter.

8) Most of the discoveries in physics like buoyancy, gravity, electromagnetism, atomic theory of Dalton, gas laws, laws of thermodynamics, the periodic table of elements, and so on were discovered when there

were no computers, satellites, powerful telescopes, and a host of other sophisticated equipments and instruments, which are currently available. However, even with this immense technical advancement, it is strange that the mystery of gravity, missing antimatter, and a host of other such important aspects of fundamental physics remain unresolved. I was wondering about the reasons behind this, when I came across the following quote by Richard Feynman, which hits the nail neatly on its head. *"In this age of specialization men who thoroughly know one field are often incompetent to discuss another. The great problems of the relationships between one and another aspect of human activity have for this reason been discussed less and less in public."* These words are not only wise but also true. When we peer into a powerful microscope, we lose perspective of the bigger world, and likewise, when we see through a powerful telescope, we lose perspective of the smaller world. The trick is to always remain grounded; connect the micro and the macro to the real world that we live in. Although the earlier scientists and physicists were at a disadvantage due to lack modern equipments, they had a different kind of advantage; most of them researched in more than one branch of physics and hence they were more adept at linking the various branches together due to which they had a broader perspective of the natural world. They made maximum use of brainpower and most of them were natural philosophers. On the contrary, today the emphasis is almost entirely on the data collected from various sophisticated instruments. Although the data collected from various experiments are definitely of vital importance in further improving our knowledge of the universe, we should not forget that true knowledge ultimately lies and emerges from the human brain. No instrument, however sophisticated, can substitute it as the knowledge for making the instruments has also emerged from someone's brain. One of the finest examples of brainpower is Albert Einstein.

9) I was surprised when I became aware that even the scientific community is dogged by politics, favoritism, egos, and so on due to which they are divided into various groups. The majority group is known as the mainstream and the others are called as minority or fringe groups. The fringe groups who are also represented by respected and brilliant scientists, are not only ridiculed for their ideas, but their articles and

research papers are not published in reputed science journals as they dare to question the veracity of certain theories pertaining to relativity, ether, and so on. In other words, there is an unwritten ban on scientists, physicists, and others whose ideas and theories do not conform to the mainstream views. This reminds us of the times when it was a crime to talk of the heliocentric model of the solar system.

In my quest for truth, I was shocked when I came across various theories, which I had earlier thought were absolutely true now sounded highly illogical. It was as if I had blinkers on my eyes all these years and someone had just removed them. **For the first time in my life, I was thinking in a highly independent and questioning manner and not just mindlessly accepting everything that I had been taught and had studied.**

I was an avid reader of books and novels in my younger days, but had never dreamt even for a single moment, nor did I have any desire or intention whatsoever, to pen a book on any topic or subject. However, because of various ideas and theories that kept popping up in my mind, I am compelled to share my thoughts with the scientific community and the public by writing a book.

My son and a few of my friends were skeptical when they learnt that I was about to publish a science book, that too on gravity, a subject that has tested the greatest and most brilliant of minds ever born. Their skepticism arose from the fact that I was not a professor, a scientist, or a physicist. Initially even, I felt nervous and developed cold feet, but then the following thought brought my confidence back. Is it necessary or is there a rule, which says that an author who is going to write about murder mysteries, robberies, or rape must first commit these crimes? Many people in this world have taken up professions that are entirely different from what they had studied and have come out with flying colors. Many people who have not done any business management courses now own huge multinational business corporations.

Here I would like to assure my readers that although I am not academically highly qualified, I have always had a deep interest in physics. I also have vast technical knowledge and practical experience. This book is a result of my research conducted almost continuously from the past 3 years and I state with certitude that I have not put in such effort even during my school or vocational training days. Moreover, as I have stated earlier, some unexplainable phenomenon or forces have literally compelled me to undertake this research and indirectly enlightened me with the relevant information and knowledge. Otherwise, how can an ordinary person like me, who struggled to write an essay or even a letter in my school days, gain so much knowledge and write a full-fledged book on the subject of gravity and other important matters in a short span of 3 years. Although science has no answers to the questions posed at the beginning of this book, I have logically tried to answer at least some of them. However, I hasten to add that I have not had any dreams or visions where some secrets of nature were revealed to me. The following quote by Einstein aptly describes the manner in which I came across new ideas and theories during the research period. *"The intellect has little to do on the road to discovery. There comes a leap in consciousness, call it intuition or what you will, the solution comes to you and you don't know how or why."* When we read articles written by great personalities, with an open and unprejudiced mind, we not only gain more knowledge but also begin to think like them. It is almost as if we have entered their minds and have begun sharing their thoughts.

I do not claim to, nor have I discovered anything new or sensational. All the contents of this book is in part basic physics learnt in school and the rest is information, most of which pertains to statements and theories associated with eminent personalities like Aristotle, Archimedes, Rene Descartes, Galileo Galilee, Sir Isaac Newton, Leonhard Euler, Albert Einstein, Paul Dirac, and a host of others, collected from the Internet and other sources. All I have done is making use of the bits and pieces of information collected and assimilating them, akin to how we go about with a jigsaw puzzle, formed a new hypothesis in which the strong nuclear force, weak interaction, and gravity are not independent forces but are indirect forces arising from the same cause. The atom and the empty space associated with it as well as outer space are central to the new hypothesis because the universe is made of only these two entities and all forces, fundamental or otherwise, should arise only from them.

Although personally, I firmly believe in the hypotheses proposed in this book, they may sound unbelievable to someone who is reading it for the first time because some of the ideas expressed in it are totally opposite to what we now believe to be true. That is why I again reiterate that new hypotheses posited in this book are based on the theories propounded by the great personalities whose names I have already mentioned. They also answer and address every aspect of gravity and a host of other things in a logical and holistic manner. Moreover, several physical evidence exists, which supports the new hypotheses. However, even if the evidence turns out to be insufficient and does not withstand scientific scrutiny, it will still make for very interesting and thought-provoking reading.

Hence, I request everyone to read this book and give your judgment.

Real or fiction

Bengaluru H.V. Mohanlal

CHAPTER ONE

1
Introduction to false facts (misconceptions)

"False facts are highly injurious to the progress of science for they often endure long."

Charles Darwin

There are two great mysteries existing in the world of physics namely the exact source and functioning of gravity and the whereabouts of the supposedly missing antimatter. Although this book is mainly about gravity and the missing antimatter, before we directly go to the subject we have to discuss a little about outer space and matter, that is, atoms, because the entire universe comprises of only these two entities and the gravitational force is, undoubtedly, linked to them.

During my research, I realized that certain theories concerning space, atoms, and so on that have a bearing on the gravitational force are actually grave misconceptions, which can also be referred to as false facts. Hence, in this book, each chapter deals with a specific misconception. The first misconception gives rise to the second and the second to the third and so on. After explaining, with proof wherever possible, why I consider them to be misconceptions, I have furnished alternate theories backed by irrefutable proof.

First, let us start by finding out what exactly is a false fact or misconception from the point of physics. A false fact is a scientific theory that has been accepted to be absolutely true by the entire scientific community and is also taught in educational institutions. This theory may have been in vogue from a few years to hundreds or even thousands of years after its inception or acceptance. It is for this reason that not a single person will doubt the veracity of these theories even if they are illogical or may strongly contradict the physical reality of our existence. Throughout the ages, there have been major misconceptions in the world of science, for example, the earth is at the center of the universe, the universe is slowing down due to gravity, and so on, that later on turned out to be the exact opposite. From history, it is evident that even today we must be harboring certain misconceptions or false facts, which we are not aware of because of the natural human tendency to unquestioningly believe in authority and the relevant scientific establishments. The following quote by Albert Einstein aptly highlights this fact, *"Unthinking respect for authority is the greatest enemy of truth."* In his later years, he lamented about the ironical fact that he, who all his life had vociferously questioned authority, was himself being considered as the ultimate authority. The following are his exact words, *"To punish me for my contempt of authority, fate has made me an authority myself."*

List of major misconceptions

1) Outer space is just an empty vacuum.
2) Light requires a medium, like the luminiferous ether, to travel in.
3) Matter is a condensate of only electromagnetic energy.
4) The atom consists of 99.999% of empty space.
5) Gravity is an independent attractive force.
6) Antimatter has vanished from our universe.
7) The sun is made up of mostly hydrogen and helium.

After reading the above list of misconceptions, I am sure that you might perhaps feel skeptical and might even be tempted to just put the book away and forget it. Before you do that, I would like to assure you that most of the hypotheses in this book are backed by irrefutable proof. I am sure that the following quote by Wilfred Trotter will definitely convince you to read this book. *"The mind likes a strange idea as little as the body likes a strange protein and resists it with similar energy. It would perhaps be too fanciful to say that a new idea is the most quickly acting antigen known to science. If we watch ourselves honestly we shall often find that we have begun to argue against a new idea even before it has been completely stated."*

To prove that misconceptions or false facts exist even in this modern age, there is one, concerning the moon, which I would like to discuss with you. Although it is not directly connected to the subject matter of this book, it will decisively prove how sometimes the official or majority view and even the most brilliant of minds can be horribly wrong. Once you read it and conduct the simple practical experiment described in it, you will be more than convinced that I am indeed speaking the truth.

2

The moon paradox

> *"Truth has to be repeated constantly, because error is also being preached all the time, and not just by a few, but by the multitude. In the press and in encyclopedias, in schools and universities, everywhere error holds sway, feeling happy and comfortable in the knowledge of having majority on its side."*
>
> Johann Wolfgang von Goethe,
> a German poet and philosopher.

Introduction

The moon is the closest celestial companion of earth and when we think in terms of cosmological distances, it is literally right beside us. However, odd as it may sound, there are two opinions among the scientific community regarding whether it rotates around its center of axis as it orbits the earth or not.

Because the same side of the moon is always facing the earth, the first thought that comes to the mind is that the moon does not rotate around its axis. However, the official view is that the moon does rotate around its axis. The time taken by the moon to complete a single rotation around its axis is exactly equal to the time it takes to complete one complete revolution around the earth or in other words, the moon is in synchronous rotation with earth. However, others believe that the moon does not rotate.

Isn't it strange that even in this age of high technological advancement, when astronomers are regularly discovering new planets, solar systems, and stars situated billions of light years away, there are two views among the scientific community. A subject which on the surface appears to be so clear-cut, simple, and obvious and which is also visible to our naked eyes? Let us study the reasons put forth by people who believe that the moon does rotate and is in synchronous rotation with the earth.

Arguments supporting rotating moon

1) In fig. 1, the moon is shown in eight different locations (A, B, C, D, E, F, G, and H) as it completes one orbit around the earth from the starting point A.

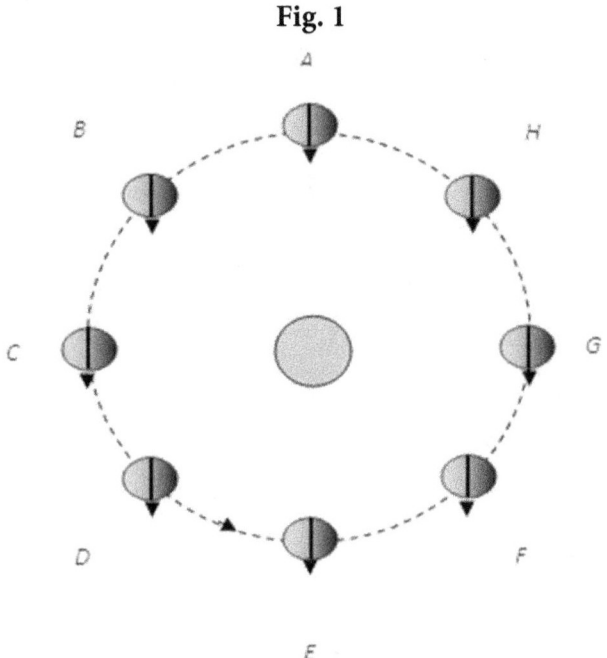

Fig. 1

The arrow marks shown just below the moon represent a fixed point on the moon's surface and the vertical line above it is shown dividing the lighter and the shaded side of the moon. If we observe the moon in the different positions, it appears that as the arrow mark is always facing downwards, that is, in the same direction, it has not rotated on its axis in the course of its orbit. Even to an observer who is positioned much farther away from the moon, the same side of the moon will always be visible. However, to an observer standing on the earth, the entire moon's surface in this instant will become visible at one point or the other as it completes one complete orbit around the earth. Hence, they conclude that since all sides of the moon are visible from the earth when it does not rotate, as it appears in fig. 1, the theory that the moon does not rotate is wrong.

2) In fig. 2, the same side of the moon is always facing the earth. Although to an observer on the earth the moon does not appear to rotate around its center of axis, to an observer positioned far away from the moon, all the sides of the moon will be visible as it completes one orbit around the earth.

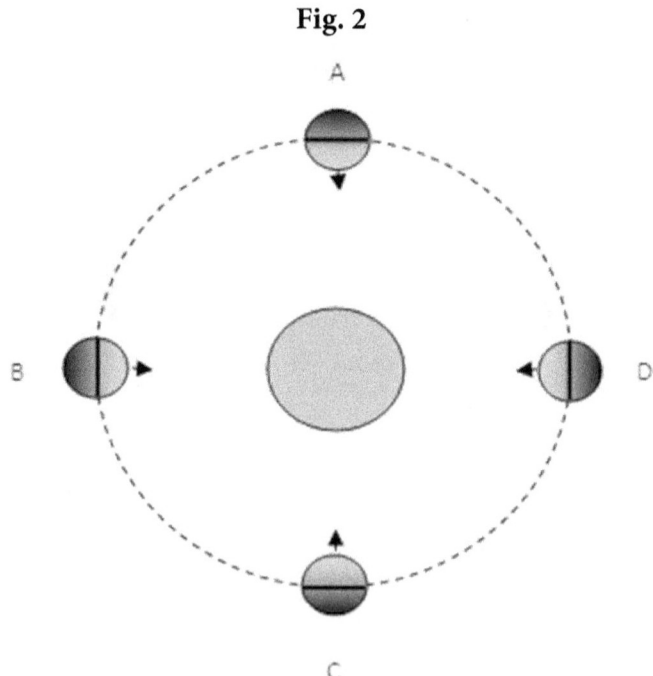

Fig. 2

At position B, it appears to have rotated by 90° with respect to position A, and at position C, it appears to have rotated by 180° with respect to position A. At position D, it would have rotated by 270° and would complete one full rotation when it again reaches position A. Hence, they conclude that the moon rotates exactly once around its center of axis for every orbit completed, or in other words, it is tidally locked with earth and is in synchronous rotation with it.

Arguments favoring a nonrotating moon

1) The moons orbit can be compared to a spherical object, which is tied to a string and being spun in a vertical plane as illustrated in fig. 3, which exactly resembles fig. 2.

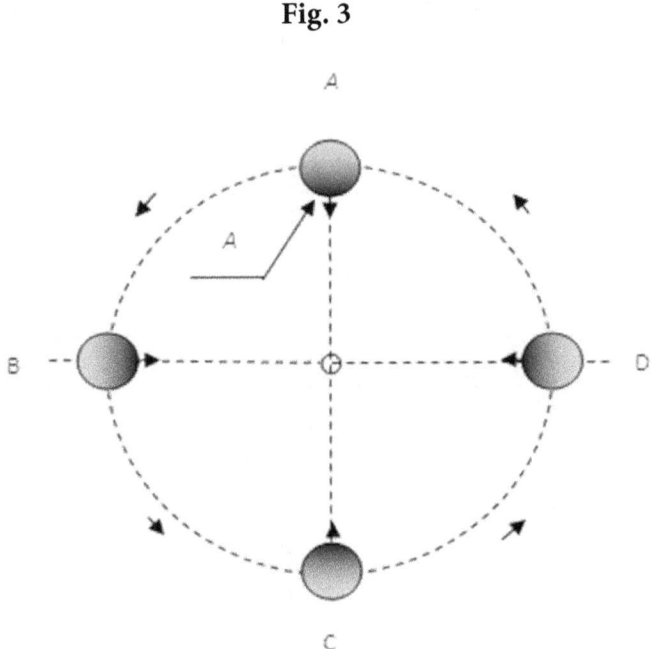

Fig. 3

As the object, which represents the moon, is always attached to the string at the same point A, it cannot rotate on its center of axis and the point of attachment A always faces the imaginary center of the circle around which it is being spun. If we for a moment visualize a rigid rod, connecting the earth and moon as in the case of the object and the string, the moon also will not be able to rotate on its center of axis. Hence, the same side of the moon will always be visible from earth as illustrated in fig. 2 although when viewed from beyond the moon's orbital path all sides of the moon will be visible.

2) The moon orbits the earth exactly in the same manner as the International Space Station ISS, space shuttle, or a satellite do. In their normal flight mode, their underbelly or the same side is always pointing toward the earth. Does this mean that they are also in synchronous rotation with the earth and are tidally locked with it? When an athlete is sprinting round and round in the racetrack, is she/he also rotating once around her/his center of axis for every round completed? Because the answer to both the questions is negative, the moon also does not rotate.

Moon with an imaginary nose and tail

Fig. 4

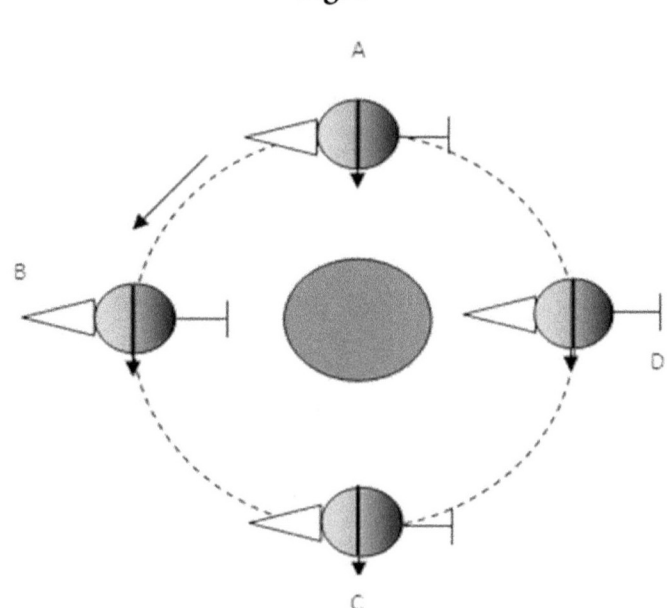

In fig. 4, the moon is depicted exactly as in fig. 1, but it is attached with a nose and a tail. In the positions A, B, C, and D, if we observe only the arrow mark and the line above it, the moon does not seem to have rotated on its center of axis. However, when we observe the imaginary nose and tail with respect to the earth, then the moon seems to be rotating in the clockwise direction. This is because at position A both the nose and tail are visible, at B only the tail is visible, at C the opposite side of the nose and tail are visible, and at D only the nose is visible.

Isn't it strange that from the perspective of the arrow and the line above it, the moon has not rotated, but from the perspective of the nose and tail, it has? From the earth, its entire surface is visible, but to an observer positioned beyond the moon's orbital path, the same side is always visible. This raises a very pertinent question as far as the method of observation is concerned. Which of the two methods is correct: observing the moon from the earth or observing it from way beyond the moon's orbital path?

Well, we now have an interesting problem because although both sides of the arguments seem logical, we know that only one of them can be correct.

Let us find out which of the two views is correct by going about in a more scientific manner because the methods and arguments adopted to decide whether the moon rotates or not seem rather crude and medieval. The method adopted should be such that there should be absolutely no room for ambiguity or for a difference of opinion to arise.

To go about scientifically, we have to correlate the number of degrees the moon rotates on its center of axis for every degree it advances in its orbital path around the earth until it completes one orbit. To find out the truth about the moon's rotation, let us conduct thought experiments in which the moon rotates four, two, one, and zero times in the anticlockwise direction in a single orbit around the earth. We are commencing from four rotations only to familiarize ourselves with a rotating moon and observe how that rotation can be represented in the sketches.

Thought experiments

1) The moon rotates four times in a single orbit around the earth

First, let us visualize that the moon is making four rotations for every orbit it makes around the earth starting from position A where the reference point is the arrow pointing downwards and is aligned with the line joining the center of the moon and the earth. To complete four rotations in a single orbit, for every degree it advances in its orbital path, it has to rotate by an angle of 4° on its center of axis in the anticlockwise direction. This means that for every 90° it traverses in its orbital path, it will complete one full rotation around its center of axis as illustrated in fig. 5.

Fig. 5

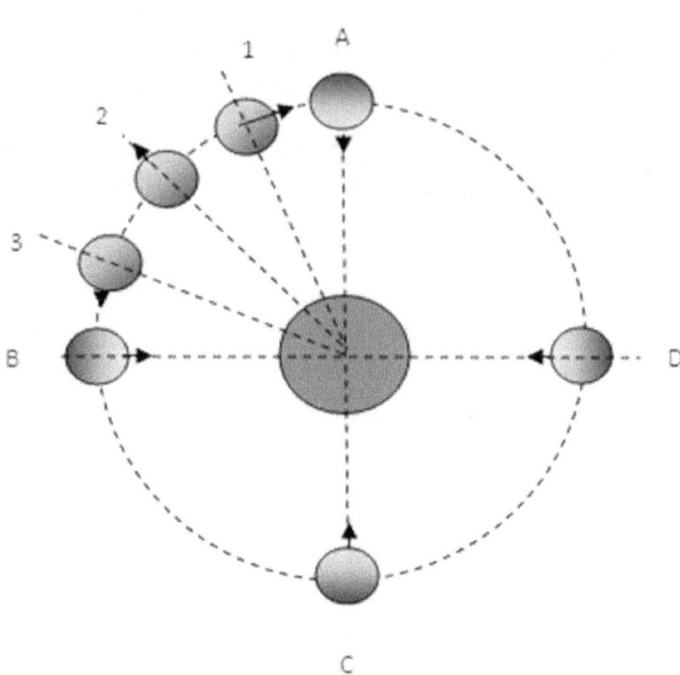

As the moon commences to orbit from position A, it will complete the first rotation in the anticlockwise direction at B. In between positions A and B, the moon is illustrated in three different positions numbered as 1, 2, and 3 to enable us to get a better idea of its rotation. From position A to 1, it traverses through 22.5° of its orbital path, and in this period, it rotates by an angle of 90° with respect to its original position A where the arrow is facing the earth. At position 2, it has traversed through 45° of its orbital path and rotated by an angle of 180 and hence the arrow is now pointing exactly in the opposite direction of the earth. At position 3, it has traversed through 67.5° of its orbital path and rotated by 270°. At position B, where it has traversed through 90° of its orbital path, it has completed its first full rotation of 360°. Similarly, it will complete the second rotation at C, third rotation at D, and the fourth rotation again at A.

An interesting aspect that we should take note of in fig. 5 is that without the moon in positions 1, 2, and 3, it is an exact replica of fig. 2 in which the moon is supposed to be in synchronous rotation around the earth.

2) The moon rotates twice in a single orbit around the earth

In fig. 6, the moon is illustrated as making two rotations in the anticlockwise direction per orbit. In this case, for every degree it advances in its orbit, it has to rotate by an angle of 2° on its center of axis. This means that for every 45° it advances in its orbital path, it has to rotate by an angle of 90°. Hence, at position B where it has traversed through 45° of its orbital path, it has rotated through an angle of 90°, and at point C where it has traversed by 90° of its orbital path, it has rotated by an angle of 180°. At this point, the opposite side of the moon will be visible from the earth. At position D, it has rotated by an angle of 225° and at E by 360°, meaning that it has completed one full rotation. The same is repeated again until it completes the second rotation at the starting point A.

Fig. 6

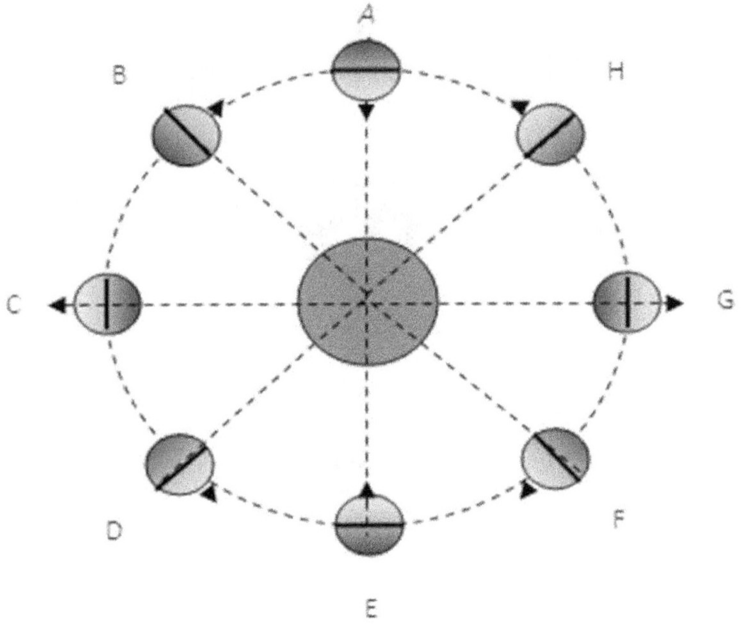

3) The moon rotates once in a single orbit around the earth

In this case, as the rotational and the orbital period of the moon is equal, for every degree the moon advances in its orbital path, it rotates by an angle of only 1° in the anticlockwise direction as illustrated in fig. 7. At position B where it has traversed through 45° of its orbital path, we can observe that the moon has rotated by an angle of 45° on its center of axis. Similarly, at C, it rotates by 90°; at D, by 135°; at E, by 180°; at F, by 225°; at G, by 270°; at H, by 315°; and at the starting point A, by 360°.

This thought experiment makes it explicitly and cogently clear that when the moon rotates even once in the anticlockwise direction, its entire surface will be visible from both the earth as well as from way beyond the moon's orbital path.

Fig. 7

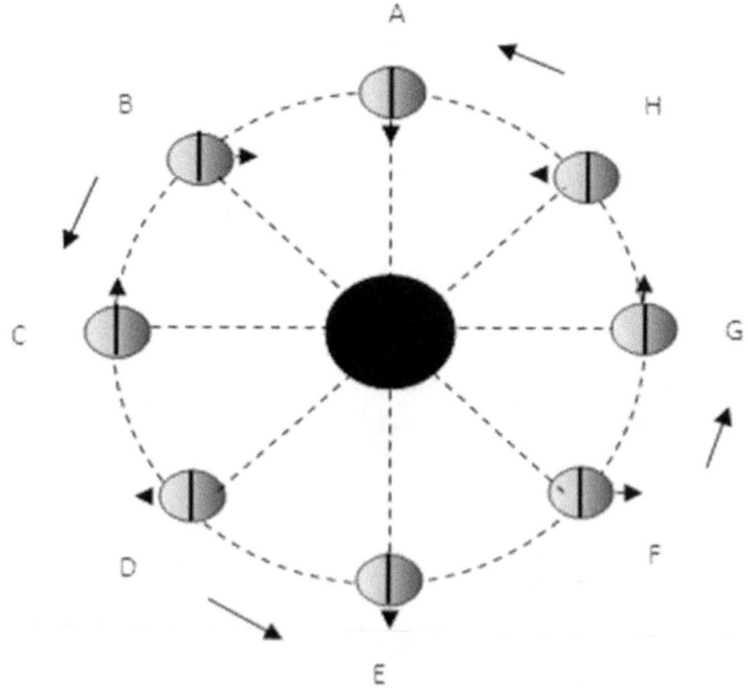

4) The moon does not rotate at all

Fig. 8 is the same as fig. 2 where it was argued that the moon rotates once. However, now we know that only if the arrow mark moves away, either in the clockwise or anticlockwise direction, from the dotted line joining the centers of the earth and the moon, can we consider that the moon has rotated. In fig. 8, as the arrow mark is aligned with the dotted line in all the positions A, B, C, D, E, and F, it can be stated with certitude that the moon in this case has completed a complete orbit without rotating even a little.

Fig. 8

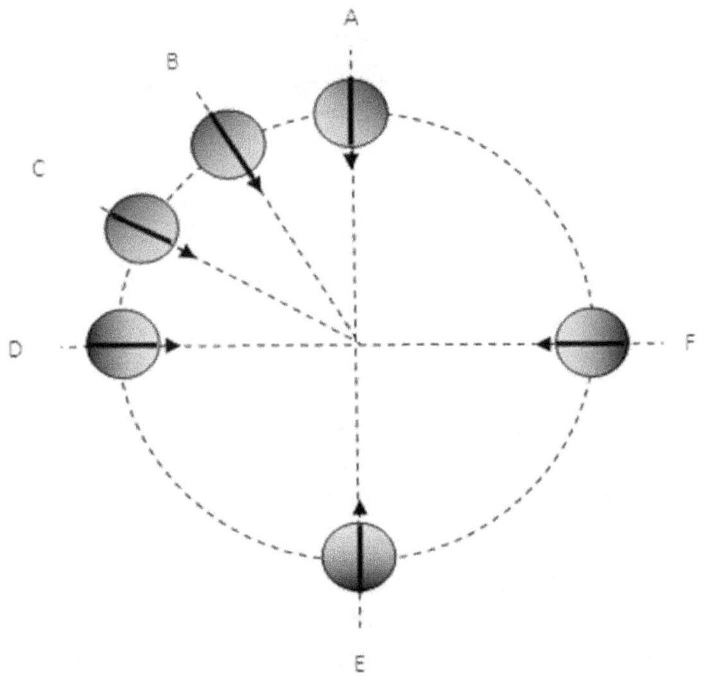

5) The moon rotates once in the clockwise direction

Let us now conduct a thought experiment in which the moon rotates once in the clockwise direction and observe the results.

In this case, for every degree the moon advances in its orbital path in the anticlockwise direction, it rotates by an angle of 1° in the clockwise direction as illustrated in fig. 9. At position B where it has advanced by 45° in its orbital path, it has also rotated by an angle of 45° on its center of axis. Hence, at position C, it would have rotated by an angle of 90°; at D, 135°; at E, 180°; at F, 225°; at G, 270°; at H, 315°; and at the starting point A, 360°.

Fig. 9

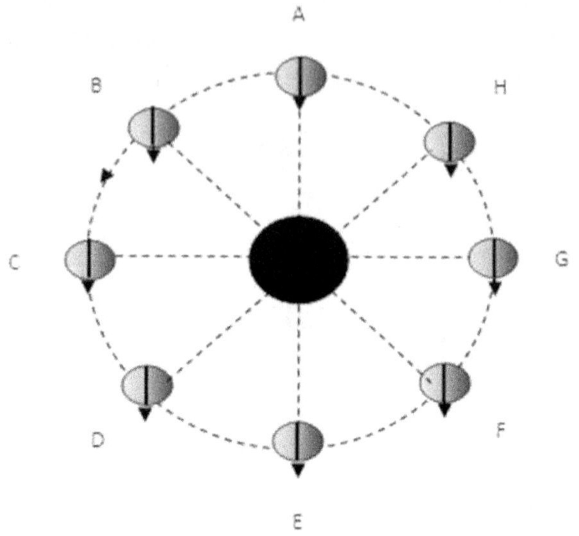

The most interesting thing that we can observe in fig. 9 is that throughout the orbital path of the moon, that is, from start to finish, the reference point on the moon, which is the arrow, always points downwards. Hence, from an overall perspective and also when viewed from outside the orbital path of the moon, it appears that the moon is not rotating at all when in fact it is rotating by an angle of 1° per 1° advancement in its orbital path. That is why the entire surface of the moon will be visible from the earth in this instance. This equal degree of revolution in the anticlockwise direction and rotation in the clockwise direction gives rise to the illusion that the moon is not rotating at all.

Conclusion

Our thought experiments prove beyond any doubt that the moon orbits the earth without rotating by even a single degree. It also highlights the fact that if the moon were to rotate even once in a single orbit around the earth either in the clockwise or anticlockwise direction, as illustrated in figs. 7 and 9, its entire surface will be visible from the surface of the earth. It is only when it rotates once in the clockwise direction that the illusion that it is not rotating is created, that too for an observer positioned way behind the moon's orbital path.

Other methods of verification

1) Animation

People who have knowledge of computers and animation can easily program the moon rotating at different speeds as it orbits the earth in an anticlockwise direction. They can observe and verify if the results match with what is stated in this book.

2) Practical experiment

We can assemble a simple model of the moon orbiting the earth with the help of a few wooden pieces and nails or with cardboard and a few ball pins as illustrated in the figures below.

1) A flat wooden plank as illustrated in fig. 10, which is in the shape of a square measuring about 300 mm per side on which the other two parts can be assembled.

Fig. 10

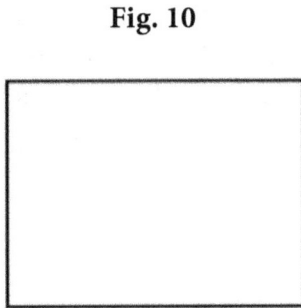

2) A flat wooden piece having a thickness of about 5 mm, width 20 mm, and length 250 mm with one hole of diameter 3 mm at one end and a dotted line drawn on it as shown in the figure.

Fig. 11

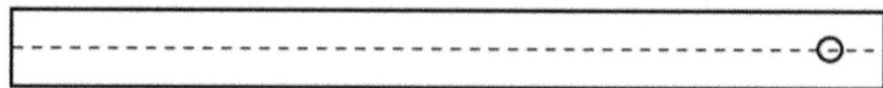

3) A round wooden bush measuring 25 mm in diameter and thickness 5 mm. It should also have a tiny hole in its center and an arrow mark drawn on it as illustrated in fig. 12 to act as a reference point. This bush will represent the moon in this experiment.

Fig. 12

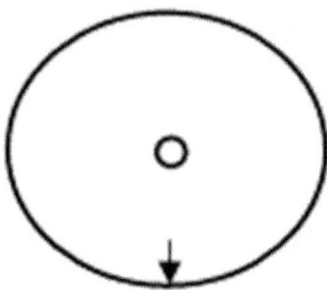

Assembly and experiment

Next, let us assemble the three objects as illustrated in fig. 13 with the help of two steel nails. The bush with the arrow is nailed to the flat bar and the bar itself is nailed to the square board at the center. The circular bush represents the moon and the intersection of the dotted lines represents the earth's center.

Fig. 13

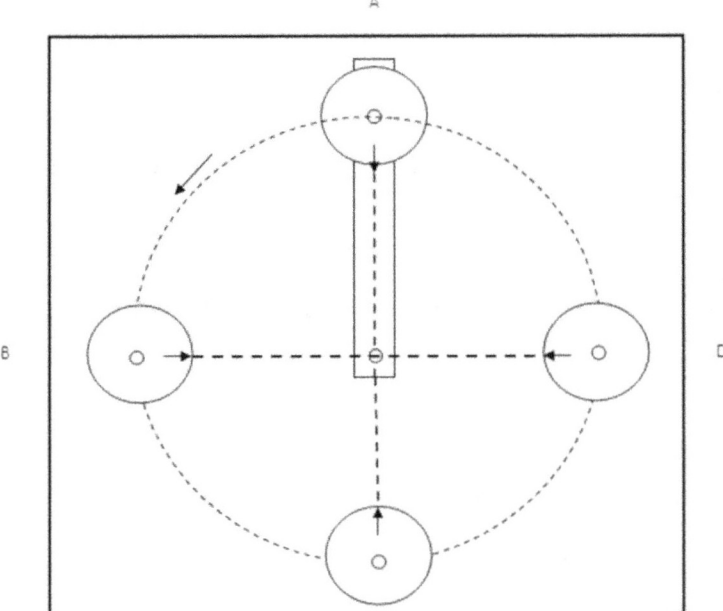

If the arrow mark on the bush deviates away from the dotted line, then we know that the bush (moon) is rotating. If the bush is rotated in the anticlockwise direction until the arrow mark comes back to the starting point, then it has completed one rotation around its center of axis. Similarly, it can be rotated in the clockwise direction also. However, there is one very important point that we should keep in mind which is that irrespective of whether the bush is in position A, B, C, or D when it is moving around the earth, only if the arrow mark deviates away from the dotted line can we say that the bush representing the moon has rotated. For example in fig. 13, the arm pivoted to the board is shown moving in the anticlockwise direction from position A to B, C, D and again back to A with the arrow mark always aligned to the dotted line. Decide for yourself as to whether the bush has rotated or not in this instance and conduct all the above-mentioned thought experiments with the help of this simple contraption and draw your own conclusions.

Moral

> *Unthinking respect for authority is the greatest enemy of truth.*
>
> Albert Einstein

Because sometimes

> *The exact contrary of what is generally believed is most often the truth.*
>
> Jean De La Bruyere

This chapter on the moon also illustrates how, sometimes,

- the authority can be wrong,
- the majority can be wrong,
- the most brilliant of minds can also be wrong, and
- how conformity wins at the cost of common sense and logic.

Discussion on Einstein's lecture on the topic of ether and relativity

Before we begin our discourse on misconceptions, it is absolutely necessary to read the entire contents of Albert Einstein's lecture presented on May 5, 1920 in Leiden (Holland) on the subject of relativity and ether. This is because it contains invaluable information along with the thoughts and ideas prevailing at that time on the subjects of space, matter, energy, gravity, theory of everything (TOE), and a host of other important issues. As I was new to this subject, I read it a number of times and each time I discovered something new. In fact, the contents of this lecture proved invaluable in compiling this book. Methuen & Co. Ltd. London has published a translated version in 1922. Just for information sake, I would like to mention here that most of today's scientific community is not aware of the contents of this lecture and even those who are aware completely ignore it. Strangely, Roland Clark has not even made a mention about this lecture in Einstein's biography.

For those of you who are wondering about ether, it would suffice, at present, to know that it is an ethereal substance believed to pervade the whole of space from thousands of years. We will discuss more about it after reading the contents of the lecture.

The lecture can be accessed from the Internet and is titled as Ether and Relativity by Alfred Einstein.

General observations from Einstein's lecture

The lecture throws light on how major decisions, concerning fundamental physics, evolve based on opinions, hypotheses, and theories of certain individuals, which later on become sacrosanct facts for future generations. The following quote by Einstein aptly describes this process, *"Concepts which have proved useful for ordering things easily assume so great an authority over us, that we forget their terrestrial origin and accept them as unalterable facts. They then become labelled as 'conceptual necessities,' etc. The road of scientific progress is frequently blocked for long periods by such errors."*

Examples: Because ether is not detected, it is assumed that it does not exist, and as a result, the mechanical views of nature are gradually reduced to that of electromagnetism. However, the fact remains that Maxwell's equations of electromagnetism cannot be used to explain or calculate mechanical phenomena for which we have to use Newton's laws of motion and vice versa.

As regards the ether, Einstein makes it explicitly clear that although he discounted its necessity in the special theory of relativity published in 1905 he was compelled, for the strong reasons mentioned in his lecture, to again bring it back to life. In fact, he concludes his speech by saying that space without ether is unthinkable. Moreover, not once during the entire lecture does he suggest that the warping of space-time endowed physical properties to space and caused it to curve. As he is the person who wrote both the special as well as the general theory of relativity, he definitely knew what he was adducing to in his lecture. Hence, he should be considered as the ultimate authority on the subject and his statements contained in the lecture should be treated as facts.

However, in gross violation of Einstein's postulates made in his lecture, there is an unwritten ban on ether theories from the past 100 years even until now due to which it is taboo to most contemporary scientists. In fact, most scientists even fear to talk about it. Isn't it truly strange that the scientific establishment, which considers Einstein as the ultimate genius, ridicules and disregards his views on ether, a subject on which he firmly believed in, was absolutely sure about and also had the fullest authority to comment upon? After all, there must have been a very valid and compelling reason for him to make such a statement, which is contrary to his earlier views.

By banning ether, not only has a great injustice been done to Einstein and the numerous illustrious ether believers of yesteryears but also to the present generation of scientists who now have to work with only one component or entity, that is, matter. This has not only created a huge void in the world of fundamental physics but is also the root cause behind the failure to resolve the mystery behind the exact source and functioning of gravity and the whereabouts of the missing antimatter, which we can also simply refer to as the second matter.

Important points to be noted from Einstein's lecture

1) An ethereal substance called ether pervades the continuum of space.

2) Physicists believe that this ether is a different kind of matter and it exists even in gross bodies, that is, ordinary matter.

2) Space is neither homogenous nor isotropic implying that its density is not uniform. This conforms to Newton's later postulate regarding the gradient-density ether.

4) Space is associated with the gravitational field, whereas matter is associated with the electromagnetic field.

5) Nature might have endowed the gravitational ether with fields of quite another type, for example, with fields of a scalar potential instead of fields of the electromagnetic type. May be that is why scientists are unable to detect the gravitational field with instruments designed to detect and measure the electromagnetic fields.

6) Although, according to the general theory of relativity, gravity is the curvature of space-time, Einstein makes it amply and explicitly clear, that it is the presence of ether, which endows space with physical properties.

7) Matter contained in objects is not only the bearer of velocities, kinetic energy, and mechanical pressures, but also of electromagnetic fields, clearly establishing its dual nature.

8) As physicists were unable to detect the ether, they assumed that it did not exist.
9) Physicists gradually abandoned the mechanical views of nature thereby reducing the principle of mechanics to that of electricity.

Conclusion

Point 8 and 9 highlight the fact that whenever great discoveries are made, in science, there is a paradigm shift in scientific thought and theories. When these paradigm shifts took place initially, there was opposition to the new ideas and theories. However, with the passage of time, they gradually met with more and more acceptance not because everyone accepted the new theories but because the number of opponents continuously decreased due to death and so on. The following quote by Max Planck aptly highlights this phenomenon: *"A new scientific truth does not triumph by convincing its opponents and making them see light, but rather because its opponents eventually die, and a new generation grows up that is familiar with it."*

However, this sometimes leads to an unintended crisis of monumental proportions if that supposed truth is not entirely or is only partially true. Then a whole generation grows up believing that something untrue or only partially true, is the absolute truth.

Einstein devoted the latter part of his life to develop a TOE, referred as the "holy grail" in modern physics, by combining both the gravitational and the electromagnetic fields. However, he was unsuccessful in his noble endeavor.

4
ETHER

1) Introduction

Right from the time of Aristotle, most eminent and renowned scientists and scholars of yesteryears like Rene Descartes, Sir Isaac Newton, Leonhard Euler, Maxwell, Eugene, Kelvin, Fitzgerald, Henry Poincare, J.J. Thomson, Fresnel, Hedrick Lorentz, Einstein, Nikola Tesla, Paul Dirac, and a host of others not only firmly believed that space was filled with ether but that it was also a constituent of matter. Many also believed that light and the gravitational force were able to propagate through space because of the existence of this medium in it. There is another reason why space cannot be just an empty vacuum.

Living creatures are able to exist and move on the earth's surface because of the presence of soil, which is a solid medium. Marine life exists and moves about in the medium of water, which is a liquid medium. Similarly, birds and aircrafts are able to fly because of the presence of the gaseous medium called air. From these examples, it is obvious that without the presence of a medium, controlled movement is impossible. Because celestial bodies move around in their orbits in a highly controlled manner, even space must be filled with some sort of medium, which might be similar to air but of extremely low density.

Now coming back to the gravitational ether, most people believed that it had mechanical properties like elasticity and so on. However, Hedrick Lorentz in his theory now known as Lorentz ether theory reduced the mechanical nature of the gravitational ether to that of electromagnetic to do away with the dualistic nature of matter.

At the same time, that is, in the early twentieth century, the various ether theories gradually began to lose favor because the numerous experiments conducted with the aim of detecting the earth's motion through the ether (ether drag) met with negative results. Hence, Einstein questioned the necessity of ether in his special theory of relativity. From then on, outer space became synonymous with an empty vacuum, devoid of any medium, and the scientific and subsequently educational institutions of the time discarded the various ether theories as unnecessary and irrelevant.

It is a fact that Einstein had questioned the existence of ether in his special theory of relativity; in response to his critics, he had once even stated that if physicists' ever detected ether, then his theory of relativity is wrong. However, in spite of this statement, in his 1920 lecture, he again strongly advocates the presence of ether as the following quote reveals. *"But on the other hand there is a weighty argument to be adduced in favor of the ether hypotheses. To deny the ether is ultimately to assume that empty space has no physical qualities whatever. The fundamental fact of mechanics does not harmonize with this view."*

We have to remember that failure to detect something is not the same as proving that it does not exist and ignoring something does not make it to go away or vanish.

5
Quotes on ether

The following quotes are worthy of note:

a)

> *"Nature abhors a vacuum."*
>
> Aristotle

According to the ancient Greeks, all celestial bodies floated in a circumambient substance (ether) of infinite extent. This "surrounding" was of a pure and higher nature, everlasting, alive and intelligent, in fact, divine.

b)

> *"The material Universe is purely made out of ether."*
>
> Rene Descartes

He also referred to ether as the second matter or element.

c)

> *"And first I suppose that there is diffused through all places an ethereal substance capable of contraction and dilation, strongly elastic, and in word, much like air in all respects, but far more subtle. I suppose that this ether pervades all gross bodies. Is not this medium much denser within the dense bodies of the sun, stars, planets and comets than in the empty celestial space between them?"*
>
> Sir Isaac Newton

d)

"*The whole of space that is left between the coarser bodies that are accessible to our senses, is filled with the subtle matter considered above, which therefore is called the ether or the subtle celestial air.*"

Leonhard Euler

e)

"*Whatever difficulties we may have in forming a consistent idea of the constitution of ether, there can be no doubt that the interplanetary and interstellar spaces are not empty, but are occupied by a material substance or a body, which is certainly the largest, and probably the most uniform body of which we have any knowledge. Whether the vast homogenous expanse of isotropic matter (ether) is fitted not only to be a medium of physical interaction between distance bodies, and to fulfil other physical functions, of which perhaps, we have yet no conception, but also to constitute the material organisms of beings exercising function of life and mind as high or higher than ours are at present, is a question far transcending the limits of physical speculation.*"

James Clerk Maxwell

Maxwell was a firm believer in ether and strongly defended ether theories.

f)

"*There manifests in the fully developed being Man, a desire mysterious, inscrutable and irresistible: to imitate nature, to create, to work himself the wonders he perceives. Long ago he recognized that all perceptible matter comes from a primary substance, or tenuity beyond conception, filling all space, the Akasha or the luminiferous ether which is acted upon by the life giving Prana or creative force, calling into existence, in never ending cycles all things and phenomena. The primary substance, thrown into infinitesimal whirls of prodigious velocity, becomes gross matter; the force subsiding, the motion ceases and matter disappears, reverting to the primary substance.*"

Nikola Tesla

g)

"*The matter of which I have been speaking so far is the material which builds up the earth, the sun, and the stars, the matter studied by the chemist, and which he can represent by a formula; this matter occupies, however, but an insignificant fraction of the Universe, it forms minute islands in the great ocean of the ether, the substance with which the whole Universe is fitted.*"

J.J. Thomson

h)

"*With the New Theory of Electrodynamics we are rather forced to have ether.*"

Paul Dirac

i)

"Relativity actually says nothing about the existence or nonexistence of matter (ether) pervading the Universe, only that such matter must have relativistic symmetry. It turns out that such matter exists. About the time that relativity was being accepted, studies of radioactivity began showing that the empty vacuum of space had spectroscopic structure similar to that of ordinary quantum solids and liquids. Subsequent studies with large particle accelerators have led us to understand that space is more like a piece of window glass than ideal Newtonian emptiness. It is fitted with 'stuff' that is normally transparent but can be made visible by hitting it sufficiently hard to knock out a part. The modern concept of the vacuum of space, confirmed everyday by experiment is a relativistic ether. But we do not call it this because it is taboo."

Robert B. Laughlin (Noble Laureate in physics and presently professor in Stanford University)

k)

"There are many things that we now collectively know that tell us things about the ether. We do not recognize them though. We have not been allowed to talk about the ether in the proper 'journals,' so we have not been able to communicate. This absurd situation has been going on most of this century, so that we have been completely wasting time, mental effort and the chance for everyone

to feel one with the Universe. I am very much against the idea that understanding of the Universe should be the prerogative of an elite. The evidence is that mathematics in fundamental physics is largely counterproductive; it confuses and encourages people to try for the PhD's by adding to the tangled web of self-sustaining intellectual fantasy, without gaining any deeper knowledge of the wonderful intricacy, fantastic forces that create and organize matter, with a continuous gradation for elementary fluctuations of some basic 'phi' to galaxies, life, and our ability to wonder at all this."

Caroline Thompson

I)

"Ether is the basic substratum of all space; ether is the raw essence of the Universe. Ether permeates the innermost recesses of all matter. Without it the Universe is contrary to nature, contrary to reason and common sense. Without it the Universe is utterly absurd."

Dr. Conrad Ranzan

Conclusion

Caroline Thomson's statement is 100% true because to arrive at a meaningful formula or equation, one must first have a sound theory. Take for example gravity. Newton and Einstein's equations on gravity are more than sufficient for all the necessary calculations. However, what is lacking is the theory behind it. How exactly does mass give rise to gravitational fields and the manner in which gravity functions. The vast majority of public are only interested in sound theories, which they can easily understand more than equations. Only scientists, professors, researchers, and other professionals need equations and formulae to discharge their professional duties.

As can be gleaned from the various quotations, most of the illustrious personalities from the past and a few contemporary scientists and physicists believe that, ether, which according to them pervades the whole of space, is the primary constituent of matter contained in celestial bodies. We now know that the matter present in an atom's nucleus is a condensate of electromagnetic energy. As the visible matter is nothing but a huge conglomeration of atoms, it goes without saying that they should also contain some amount of ether or the second matter or element in a condensed form.

Regarding the ether theories, it is immaterial whether we agree or disagree about them. However, there is one thing I am sure that everyone of us will definitely agree upon, which is, the eminent persons beginning from Newton to Einstein and Paul Dirac were brilliant scientists and physicists whose beliefs cannot be just brushed aside because their various theories and discoveries form the very foundation of modern-day physics.

Hence, it is not only the bounden duty of present-day scientists, physicists, and astrophysicists, but they also owe it to these great personalities to discuss and conduct research in an unprejudiced and fearless manner as we now have easy access to space, which was unthinkable in their times. This will establish finally whether space is truly empty or an ethereal medium pervades it. If it is found that even after conducting the necessary research, including suitable experiments, that space is just an empty vacuum, then the ether theories will die a natural death and it will no more be necessary to ban them.

However, if on the other hand scientists discover that an ethereal medium is present in the continuum of space, then it will not only have a profound impact on fundamental physics, but will also give rise to a new paradigm as far as our understanding of nature is concerned.

CHAPTER TWO

1
Misconception number 1

"Nature never deceives us; it is we who deceive ourselves."

Jean Jacques Rousseau

Outer space is just an empty vacuum

Introduction

We have already discussed about how the establishment discredited and discarded the ether theories in the early twentieth century. However, we should take note of a strange and ironical coincidence, which is, that as long as the ether theories were in vogue and intensely debated, humans had no access to space and hence could not practically verify its physical properties or true nature. However, just a few years after discarding the ether theories as unnecessary, humankind conquered space and now has easy access to it. It almost seems that nature does not want humans to discover her well-kept secrets all at once.

Although the contemporary scientific community does not believe that space is pervaded with ether, they believe that it is neither empty. According to quantum theory, the vacuum of space is full of fundamental matter particles like quarks called virtual particles, which are always paired up with their appropriate antiparticle counterpart. They pop into existence and almost immediately collide annihilating each other. However, there is no clarity with respect to their emergence and their purpose of existence in outer space.

As we now have access to space, let us try to find out if space is physically empty or pervaded with an invisible substance or medium. However, let us desist from referring to the substance that we are in quest of as ether or pether because, as we all know, the name by itself is not important and implies nothing.

Surprisingly for achieving this task, we will neither require any instruments like the interferometer or spectrometer, nor will we have to conduct any experiments in space because space agencies like National Aeronautics and Space Association (NASA), European Space Agency (ESA), and others have already, knowingly or unknowingly, accomplished this task in the course of their numerous journeys into outer space. All we have to do is keenly observe the various phenomena that occur both inside and outside spacecrafts, like the ISS and shuttle, which are in outer space and then decide if space is really empty or not. For this I assure you, we do not require the analytical or observational skills of Sherlock Holmes.

Beginning of outer space

To study the various phenomena, we have to know from where space actually begins. Although there is a vast difference between the conditions existing in space and in a planet's atmosphere, it is widely believed that there is no clear boundary between the two because as the altitude increases the density of air decreases exponentially until it almost ceases to exist above 100-km altitude. In other words, from the perspective of air pressure or density, it is not clear at what distance from the earth's surface space starts or the atmosphere ends. However, for practical reasons, the ESA considers that space begins from 100 km above the earth's surface, also known as the Karman line, whereas NASA has fixed it as 80 km.

Phenomena that occur only in outer space
1) Weightlessness
Introduction

It is normally thought that any object that enters space, that is, rises above 100-km altitude from the earth's surface, experiences weightlessness, which is attributed to microgravity. However, this is not entirely true because objects and astronauts experience weightlessness only when they are inside a spacecraft, which is in space. The spacecraft itself will fall back to the earth if it does not possess the required orbital speed. Similarly, even astronauts will fall back toward the earth if they happen to step out of a sounding rocket or a spacecraft, which is traveling below the required orbital speed.

Various reasons attributed to weightlessness

a) There is no gravity in space as space is a vacuum. This is not true for two reasons. Gravity very much exists in space because the moon orbits the earth and the earth orbits the sun because of gravity. On the earth's surface, if we create a vacuum in a perfectly sealed room, we will notice that the objects present in the room will not float.

b) Objects inside orbiting spacecrafts, for example, ISS as well as the spacecraft itself, whose speed is about 27,000 km/hr, are in free fall around the earth's surface as in the case of a falling lift or a drop tower in an amusement park. Because people working in reputed space organizations make this statement, most of us blindly believe in what is stated. However, the following example will prove beyond any doubt that the high speed of the orbiting spacecraft is not the true reason behind weightlessness in space.

Weightlessness in a sounding rocket

Most of us know that space organizations use rockets to carry satellites into space and place them into regular orbits. However, there is another type of rocket known as the sounding rocket, which they primarily use for experimental and nowadays even tourist purposes. Private space agencies sold advance tickets at the rate of $200,000 per person to take them above 100-km altitude from the earth's surface in specially built spacecrafts, for about 5 minutes or so, during which time they would experience weightlessness. Unlike a regular rocket, the sounding rocket does not go into a regular orbit but terminates a few minutes or hours after entering space and hence its flights are called as suborbital flights.

Its flight path is shown in fig. 14 and is represented by the big arrowheads. It flies vertically upwards and its engines are switched off just before it reaches an altitude of 100 km from the earth's surface. Once it reaches the desired altitude in space, which can vary from a few hundred to thousands of kilometers, it takes a U-turn and proceeds downwards toward the earth's surface eventually landing on to it with the help of a parachute.

Fig. 14

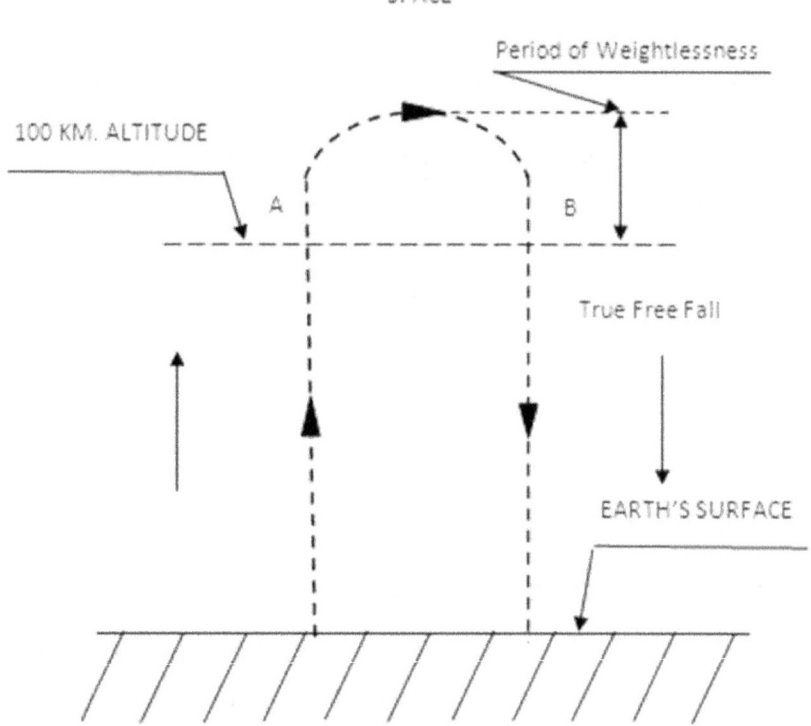

An interesting phenomenon that occurs inside a sounding rocket as it ascends above an altitude of exactly 100 km is that all the unfixed objects within it will begin to magically experience weightlessness or microgravity and start floating even as the rocket is ascending. The period of weightlessness will last until the rocket reaches maximum altitude, where its vertical velocity becomes zero, and continue until it descends and reaches an altitude of 100 km. However, the instant it descends below 100 km, weightlessness disappears abruptly and all the objects regain their weight.

However, the spacecraft always appears to be in possession of its weight, even when it is above 100 km, which is borne out by the fact that as it ascends, it decelerates until its vertical speed becomes zero after which it begins to fall down. At that altitude from the earths' surface, the spacecraft and all the objects within it should possess around 90% of their weight on the earth and hence logically they should not float but remain on their resting places inside the spacecraft. True free fall occurs when the spacecraft is below 100 km and plunges at 90° to the earth's surface. However, strangely, objects within it do not experience weightlessness during this period.

Can the motion of orbiting bodies be compared to free fall?

The concept that all orbiting bodies like satellites, moon, earth, and the other planets are in a state of constant free fall around the primary bodies is in itself a huge misconception for the reasons mentioned below.

We are all aware of the fact that although the surface of the earth is a sphere, in reality or from mother nature's perspective, it behaves exactly similar to a flat surface. Odd as it may sound, we will not be wrong if we refer to the surface of celestial bodies as spherically flat surfaces. The following examples along with the sketch illustrated in fig. 15 will highlight this fact.

Fig. 15

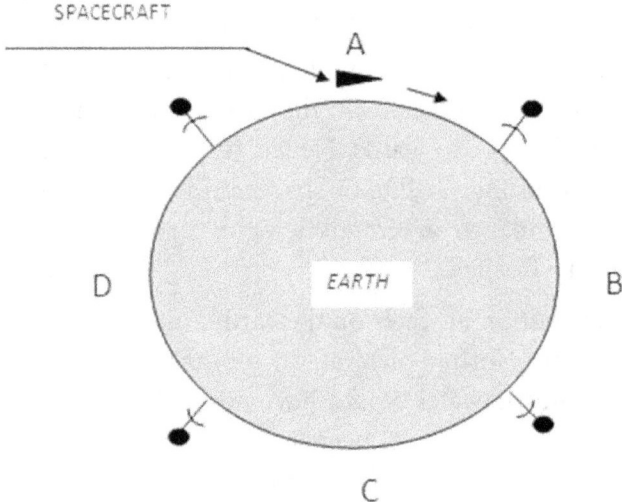

a) There are billions of people living on the earth in almost every region of the earth's surface. However, each one of us irrespective of our location always feel that we are on top, that is, at position A, and standing upright with the sky above us even though fig. 15 clearly illustrates that this is not true. In reality, we should get this feeling only when everyone is standing on a flat surface.

b) An aircraft or a spacecraft is not falling downwards when it moves from point A to point B or C and likewise not climbing when it moves from point C to again point A. At all times during the flight, the passengers and crew inside the crafts will feel that they are flying on a flat surface and in an upright position and the earth's surface is always in a downward direction, whereas the sky is always in the upward direction. It is as if the spacecraft itself is stationary and the earth along with the sky is rotating. Moreover, as the flight's path is always parallel to the earth's surface, it is not only hard but also illogical to visualize free fall in a horizontal direction. That is why when we represent the earth's surface on paper, in such instances, it should be a horizontal straight line.

c) Most human beings will be scared out of their wits when they are in actual free fall due to which there a huge rush of adrenalin because of which we cannot withstand free fall for lengthy periods. However, if we observe the videos of astronauts in the ISS, they are neither screaming nor do they appear to be scared. On the contrary, we can see them enjoying the sensation of actually being weightless as the spacecraft is flying exactly similar to an aircraft flying in the earth's atmosphere. Only when a spacecraft's distance from the earth's surface decreases, can we say that it is falling downwards. Hence, it would be more appropriate to say that an orbiting satellite or spacecraft is flying freely around the earth, exactly similar to an aircraft flying in earth's atmosphere, than to say that it is in a free fall.

d) The even distribution of water on the earth's surface also clearly proves this point. If the North Pole was on top and the South Pole at the bottom, then all the water would have flown down toward the South Pole, which of course has not occurred. If the earth's surface had been perfectly smooth, that is, without any ups or downs, then the entire earth would be uniformly submerged under the oceans waters.

These points clearly highlight the fact that although celestial bodies are spherical in shape, in reality and for all practical purposes, their surfaces should be thought of as being flat. Hence, we cannot and should not equate the motion of orbiting bodies to free fall.

Thought experiment to prove that true free fall also does not give rise to weightlessness

Let us now visualize that a person is standing on a weighing machine, which is resting on the floor of a lift of a very tall building and the lift itself is on the topmost floor. The weighing machine will indicate his weight as X kg. If the cable supporting the lift were to snap suddenly, then the lift along with everything inside will begin to fall toward the earth with an acceleration of 9.8 m/sec^2. Although the weight indicated on the weighing machine will now be zero, the person will still remain in contact with it and will not begin to float like the astronauts inside an orbiting spacecraft. Similarly, if a bucket filled with water had been inside the falling lift, it would have remained on the lift's floor. In case of a spacecraft, which is in space, the bucket along with the water would have risen up.

This thought experiment clearly illustrates the fact that free fall alone cannot give rise to weightlessness. Now we will discuss the true reason behind weightlessness, which is so ridiculously simple and glaringly obvious that you will wonder why you did not think of it before.

The real secret behind weightlessness in space

To find out why celestial bodies and objects entering space appear to become weightless when in reality we know that matter contained in them is heavy, we have to start from the time of Aristotle when two manifestations of weight were recognized:

a) Weight causes the falling of nonsupported objects.

b) Weight causes the downward pressure exerted by the object on its support, when available.

The heavenly bodies were not supported and did not fall; therefore, they were inferred by Aristotle to be weightless. From the time of Aristotle until almost the sixteenth century, the geocentric model of the cosmos was in vogue and this notion is justified. However, even after the heliocentric model superseded the earlier theory, the notion that celestial bodies are weightless has continued throughout the ages even until today. One of the reasons behind this belief stems from the fact that while calculating the orbital speed of celestial bodies, only their orbital radius is taken into account, whereas their weight and mass are not considered. If we put a golf-sized ball into orbit at the same distance as the moon, its orbital speed will be exactly equal to that of the moon.

This is a strange paradox because from the perspective of space, celestial bodies do not seem to possess weight. However, on the surface of these bodies, for example, our earth, we know that objects and matter possess weight.

This phenomenon has to do with a special characteristic of space and that is, there are no upward or downward directions from the perspective of space. Although this may sound a little strange, as we are so used to the concept of up and down on the earth's surface, it is absolutely true. In this context, let us visualize a celestial body, which is far away from the gravitational influence of all other bodies or where the gravitational forces are almost negligible as illustrated in fig. 16 and find out how this body would behave if it were in a state of absolute rest. In such a situation, it will remain suspended in the same spot in space, and under similar circumstances, all other celestial bodies will behave in the same manner irrespective of their mass. Let us now find out why they behave in this manner.

Fig. 16

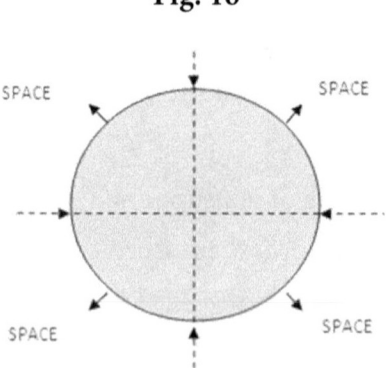

The reason why celestial bodies remain suspended in space in the absence of a gravitational force is very simple. If we consider the body in fig. 16, we will notice that from every point on the body's surface, space is always in an upward direction and downwards is always toward the center of mass of the body. Hence, weight of matter contained in the body from all directions will always act only toward the center of mass of the body and not in any other direction. Even if we presume that weight is a natural trait of matter, the same reasoning will hold well. However, when we consider the body as a whole, we will notice that it cannot move or fall in any direction because for it all directions are the same and hence it will remain floating in space even though matter contained in it is heavy.

However, if appropriate force is applied to this body from any direction, it will begin to move in a straight line in the opposite direction. This also means that inertia is felt from every conceivable direction. Astronomers have found thousands of such freely floating bodies, which do not orbit any other celestial bodies, in the empty space of our universe and have named them as rogue planets.

This is an extremely important discovery that we have made because it proves without an iota of doubt that although celestial bodies appear to be weightless from the perspective of space, they are always in possession of their weight. Even if weight is a natural trait of matter, that is, it arises from the heaviness of matter itself, celestial bodies would still appear to be weightless from the perspective of space.

Now let us assume that the gravitational field abruptly ceases to exist in the entire universe and all the orbiting heavenly bodies are deprived of their motion thereby coming to a sudden standstill and find out the consequences in such a scenario. In such an instance, all the celestial bodies will remain suspended in their respective places in space because from their perspective all directions are the same. Although this seems strange and hard to believe, because in our day-to-day lives we observe that all unsupported objects fall down, it is indeed true.

Thought experiments to prove that there are no upward and downward directions in space

In fig. 17, a celestial body E is shown floating in space surrounded in the distance with four other bigger bodies A, B, C, and D. The shaded outer portion represents their gravitational fields.

Fig. 17

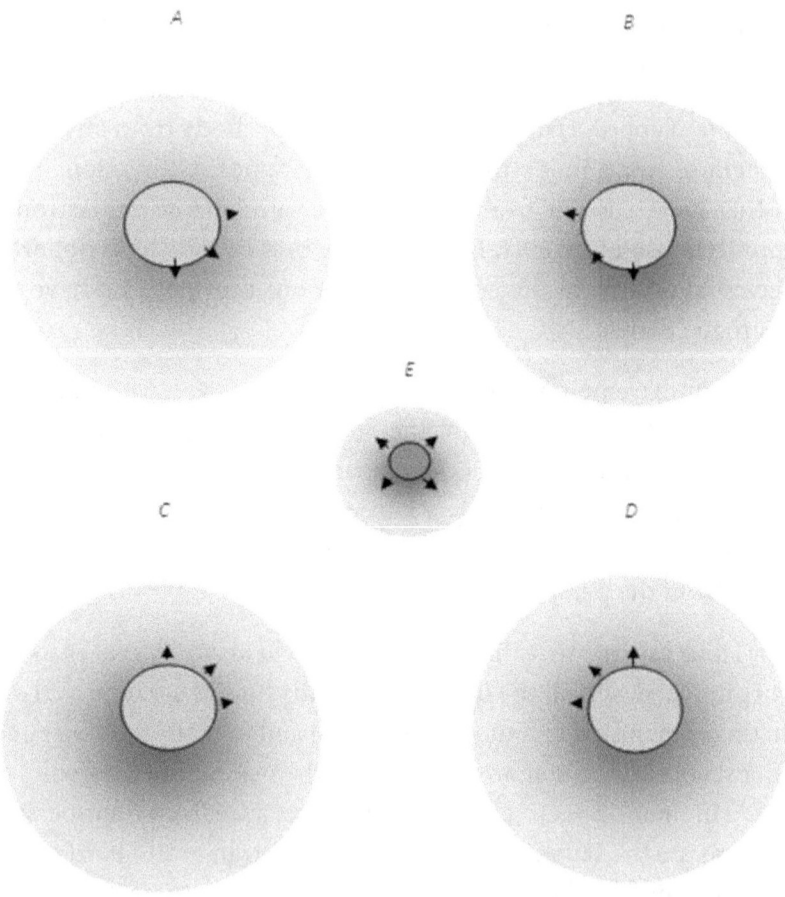

Let us presume that the mass of each of these bodies is 100 times more than the central body E, and they are so far away from it, that their gravitational fields do not influence it. In other words, the space that it is now present in is isotropic. Now let us observe what will happen if the body A moves closer to the central body E, while the others stay in their respective positions, until its gravitational field begins to act on it. The body E will begin to move or fall in the direction of A. In a similar manner, the central body E will move toward the other bodies also when they come closer to it, one at a time, though they are in different directions. In fact, E can move in every possible direction as long as there is a body with a gravitational field in that direction and it comes within this body's gravitational field. This thought experiment clearly illustrates and also establishes the fact that there are no ups and downs in space.

In the same figure, let us visualize that people are standing at the places indicated by the arrows. To people standing on the central body, all the other bodies will appear in the upward direction. Similarly, for people standing on the other four bodies, the central body will also appear to be in the upward direction although on paper it appears to be below them on top or at an angle. Similarly, from earth, the moon appears to be on top, whereas from the moon, the earth will appear to be on top.

These thought experiments clearly establish the fact that from the perspective of space, as far as celestial bodies are considered as a whole, there are no upward and downward directions. However, the concept of up and down exists only on the surface of individual celestial bodies where space is always in the upward direction and down is always toward the center of mass of the body.

Why does weightlessness occur only inside spacecrafts?

It is a well-known fact that objects and astronauts experience weightlessness only when they are inside spacecrafts. If an astronaut happened to step out of a sounding rocket which is in space, then she/he will plunge toward the earth just as the rocket will also plunge after it has reached its designated altitude. There is only one logical explanation for this strange phenomenon. As long as the astronauts are inside the spacecraft, they are shielded from coming into direct contact with the invisible and mysterious medium that is present in space by the spacecrafts body. However, as soon as they come out of the spacecraft, they fall because they are now in direct contact with the medium.

2) Spherical formation of liquid bodies

All liquids in outer space, irrespective of their mass, will immediately take on the shape of a sphere, and this phenomenon will even occur inside spacecrafts that are in space. There are two reasons attributed for this phenomenon.

a) Liquids in space take on a spherical shape because a sphere occupies the least space or volume for a given mass.
b) In liquid bodies, surface tension a property possessed by all liquids forces liquids to take on a spherical shape.

Although the first point is not a valid reason, surface tension, which plays an important role, is also not the sole reason because this phenomenon does not occur on the earth's surface or in its atmosphere. The following is the true reason behind the spherical formation of liquid bodies in space or the forces that are involved in giving rise to this phenomenon.

Mechanism by which liquid bodies attain a spherical shape in space

We are now aware that in space the weight of matter contained in any object, irrespective of whether it is in a liquid or a solid state, will always act toward its center of mass from all directions. The special characteristic of a spherical shape is that it is the only shape in which mass and the weight associated with it is uniformly distributed around the center of mass of the body. Fig. 18 illustrates an oblong body of liquid in space that later takes on a spherical shape.

Fig. 18

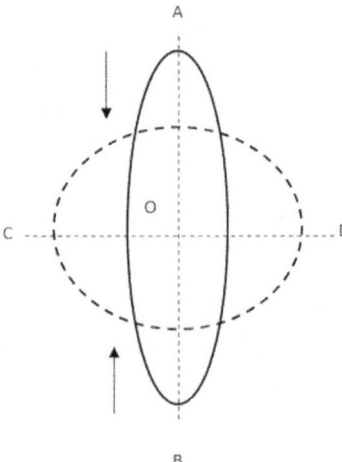

In space from all conceivable points, for example, A, B, C, and D, down is always toward the center of mass O. Hence, as there is more matter or mass across the line AB, the excess weight forces the liquid to fall toward the center of mass from both ends A and B because of which matter is pushed out across line CD. This is possible only because the molecules in the liquid body are able to freely move and slide against each other. Surface tension or the intermolecular force as usual serves the purpose of bonding the liquid molecules to each other. Matter from both the ends A and B will continue to fall toward point O until the oblong liquid body transforms into a spherical shape. Once the liquid body assumes a spherical shape, it reaches a state of equilibrium and hence will continue to retain this shape. Objects in space are also subjected to a uniform compressive force that acts from outside. I will expound on how this compressive force arises later on in this chapter.

Why liquid bodies do not take on a spherical shape on the earth

Because the earth's atmosphere is not the same as space, liquids here will not behave like independent celestial bodies but as an integral part of the earth and hence their entire weight will now act toward the center of mass of the earth. That is why, when we place even a small quantity or blob of liquid on the earth's surface, it will flatten and spread out all over instead of attaining a spherical shape because the weight of liquid present in the blob is greater than the intermolecular forces present in it. Only tiny drops of a liquid, like dew drops present on leafs in the early morning, will take on a spherical shape on the earth because the strength of the intermolecular forces in the drop are greater than the weight of the miniscule quantity of water present in it.

3) Bouncing on a vacuum

Fig. 19 illustrates the earth along with its atmosphere, whose density and pressure decreases exponentially with increase in altitude. 99.99997% of the atmospheric air is supposed to be present within 100-km altitude from the earth's surface. Hence, for all practical purposes, we can safely assume that above this altitude spacecrafts would not be subjected to any physical obstruction or resistance.

Fig. 19

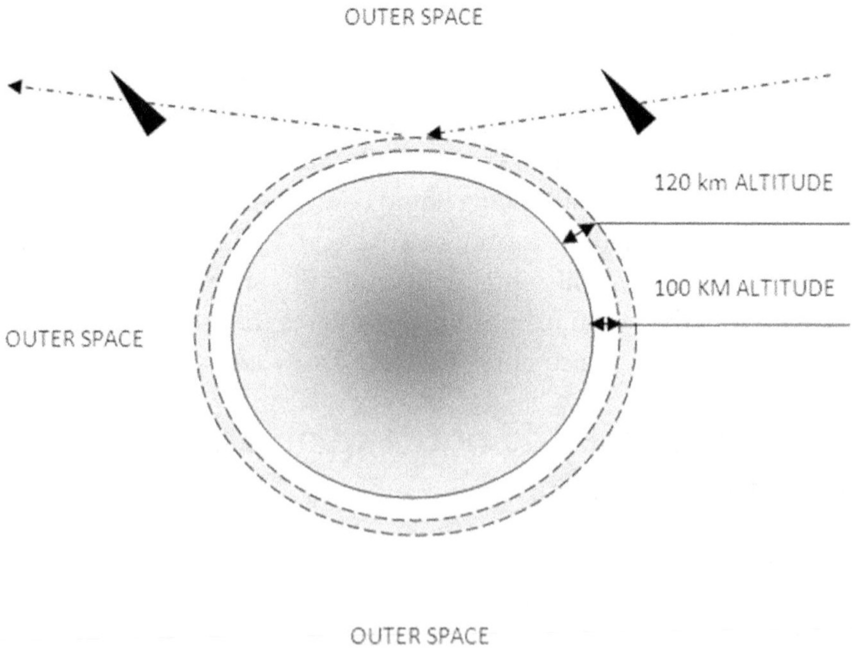

Let us now observe a spacecraft, which is re-entering the earth's atmosphere after completion of its mission. As the spacecraft, whose speed is around 27,000 km/hr, gradually descends at a shallow angle of 1° and reaches an altitude of 122–120 km from the earth's surface, it is subjected to a strange and unexplained phenomenon.

If its AOA, inclination of the spacecraft with nose up against the trajectory, is more than 40°, then the spacecraft will bounce back into space similar to how a flat stone bounces on the surface of a placid lake when thrown at a very low angle to the water surface. The fact that the massive 100-ton spacecraft so easily bounces back into space in a region which is supposed to be just an empty vacuum proves beyond an aorta of doubt that an invisible but physical elastic-like membrane is present at this altitude.

4) Heating up in a vacuum

This phenomenon is again pertaining to a re-entering spacecraft. If the AOA of the descending spacecraft is maintained at around 40°, then the spacecraft will not bounce back into space. However, as it reaches and descends below 120-km altitude, the underbelly and wing tips heat up tremendously. This sudden and intense build-up of heat is definitely not due to air because only an extremely negligible quantity of air exists at this altitude. Moreover, if the heating was really due to the presence of air, then the heating should have started gradually and increased progressively as the spacecraft descended into the increasingly dense atmosphere. However, it is an established fact that the actual heating of the spacecraft caused by atmospheric air begins only at around 74-km altitude from the earth's surface. Strangely, the spacecraft does not heat up in between 74- and 100-km altitude.

5) Orbit decay

Satellites and other spacecrafts like the ISS are subjected to severe orbit decay, which is highest just above 100-km altitude from the earth's surface and decreases gradually with increase in altitude. Small booster rockets are used to maintain the satellites in their designated orbits by firing them when required. Our moon, which is at a distance of about 300,000 km from the earth's surface, does not face this problem, as its orbit is quiet stable. It is believed that the earth's atmospheric drag causes the orbit decay of satellites in low earth orbits. However, as we know that there is only 0.00003% of air spread out in the thousands of kilometers of outer space beginning from an altitude of 100 km above the earth's surface, this does not seem to be the real reason. This proves that other than air there is not only something else pervading space, but its density also seems to be decreasing with increase in altitude, which can be deduced from the fact that orbit decay decreases with increase in altitude and vice versa. This is exactly similar to the density of earth's atmospheric air, which is densest at the earth's surface and decreases with increase in altitude.

Thought experiments that will further confirm that the environment of outer space is different from the atmosphere of celestial bodies like earth

Experiment number 1

As our earth is orbiting the sun, in the ocean of outer space, with a speed of about 30 km/sec, we can compare it as a whole to a man-made spacecraft or satellite, which is also orbiting in space. The only difference would be that the earth is orbiting the sun and the satellite is orbiting the earth. Although the spacecraft as well as the earth's atmosphere are filled with compressed air, the objects within a spacecraft will float as long as it is in space irrespective of whether it is in a regular or a suborbital flight. On the contrary, objects present within an aircraft flying in the earth's atmosphere will not float irrespective of whether the aircraft is in motion in a horizontal direction or is in a free fall. The objects within a spacecraft, which is in orbit, will also cease to float when it descends below 100-km altitude from the earth's surface even if it possesses the required orbital speed. This thought experiment proves that the environment of space is very much different from earth's atmosphere.

Experiment number 2

Let us visualize that we are standing on top of a spherical object consisting of earthly matter, having a diameter of about 4 km or more, resting on earth's surface. Let us also visualize that we are on top of this object and begin walking by trying to maintain an upright position at all times. We will notice that this task is impossible and that as we move away from the top toward the side, our center of gravity will gradually shift away from the center of our bodies, as illustrated in fig. 20 until we finally fall down.

Fig. 20

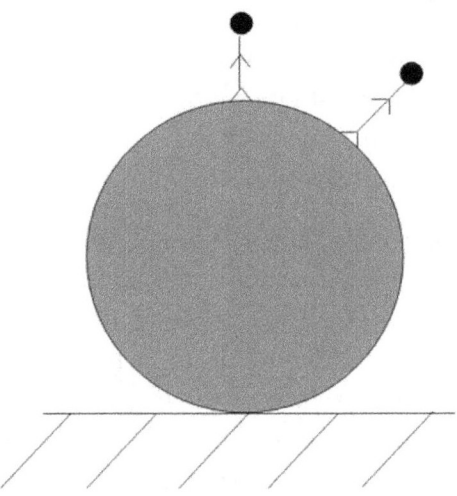

Even if we assume for a moment that if we somehow suspend it in the air, as long as it is within 100-km altitude from the earth's surface, we will not be able to walk completely around its surface. Let us now visualize that this same spherical object is orbiting the earth from a few thousand km away and find out the results. If we take into consideration the recent landing of Rosetta's Philae probe on the comet G7P/Churyumov-Gerasimenko, which is about 4 km wide, then we should be able to walk around the entire surface of the spherical object without falling toward the earth. This is because as the object is in space, it itself behaves like a celestial body and is now surrounded by its own gravitational field. Even the asteroid will lose its gravitational field if it were to land on the earth's surface.

This thought experiment clearly highlights the strange fact that as long as an object is within the boundaries of earth's atmosphere, that is, below 100-km altitude, it will not have a gravitational field of its own. However, as soon as the same object is in the environment of outer space, a gravitational field whose extent is proportional to its mass will instantly arise around it, clearly implying that outer space is pervaded with a subtle substance that gives rise to this gravitational field. The following quote by Albert Einstein lends credence to this postulate, *"the existence of the gravitational field is inseparably bound up with the existence of space."*

Why does space begin from 100-km altitude?

It is now clear that the ethereal medium that endows physical properties to outer space is present from above 100-km altitude from the earth's surface. This leads us to a very interesting as well as important question, which is, why does not this medium begin from the earth's surface or in other words, what is preventing it from coming down?

There is only one possible explanation for this phenomenon. The ethereal medium present in space and the matter contained in the earth must mutually repel each other due to which the elastic medium is pushed upwards. Because of this mutual repulsion, which is neither magnetic nor electric in nature, a compressive force will begin to act all around the earth's surface as illustrated in fig. 21. That is why, because air is also matter, it is not only prevented from escaping into space but is also compressed giving rise to atmospheric pressure. According to quantum mechanics, the gravitational force is too weak and is as good as absent at the atomic and subatomic levels.

Fig. 21

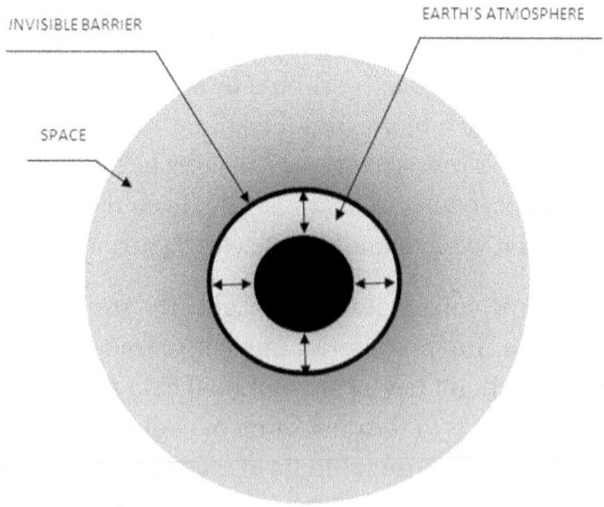

The following belief of both Newton and Immanuel Kant substantiates the new hypotheses.

In query 31 of *Optics*, Sir Isaac Newton famously advocated attraction and repulsive forces as the fundamental qualities in nature, whose causes were however unknown. Even the famous philosopher Immanuel Kant expressed the same view as is evident from his quote. *"I have applied no other forces other than attraction and repulsion to the evolution of the great order of nature: two forces which are both equally certain and equally simple."*

The following quote by Leonhard Euler, which although does not mention mutual repulsion, is uncannily similar to our postulate. *"The subtle celestial air (ether) is in a forced state, and is compressed far beyond its natural density, for which reason it exerts everywhere an unusually strong spring force and compresses all bodies."*

Conclusions

The various phenomena that we have discussed, for which there are no logical or scientific explanations, and the thought experiments clearly establish the following facts.

1) Outer space is definitely filled with a medium, which is elastic in nature and has an extremely low natural density. This assumption leads to a very important question, which is: why is friction absent in space? The answer to this question can be found by observing light, which is constituted by photons. Outer space is filled with light and other electromagnetic waves, emitted by sun and billions of other stars, which are constantly crisscrossing one another. However, it is also true that the presence of these waves does not give rise to friction. The nature of the elastic medium pervading the whole of space could be similar to the electromagnetic waves, but with one difference: the photons that constitute it do not possess any electrical charge and hence they are not easily detectable.

2) Although earth's atmosphere is full of air, the fact that weightlessness and the various other phenomena do not occur in it proves that this mysterious and ethereal substance, which is present in space, is absent here. Because it is present from above 100 km from the earth's surface, outer space also begins from that altitude.

3) An invisible but physical barrier having elastic properties exists that begins at 100-km altitude and ends at 120-km altitude from the earth's surface and whatever this barrier may be comprised of, it is definitely not atmospheric air. This invisible barrier, which is illustrated in fig. 21, is about 20 km in thickness and acts as the boundary wall that clearly separates space and the earth's atmosphere.

Its density seems to be uniform and also greater than the density of the gradient-density medium present above it. The effect of this membrane can be observed only because of the spacecraft's high speed. At lower speeds, a spacecraft will easily pass through this invisible membrane without bouncing or heating up. Even in the case of the stone, it will bounce off the water surface only when it is thrown with sufficient speed and a very shallow angle.

Max Planck, a German scientist and one of the pioneers of quantum mechanics, proposed the following when he was discussing certain aspects of stellar aberration with Hedrick Lorentz in 1899. *"The ether was compressible and accumulated with greater density around large celestial objects."* Hedrick Lorentz admitted that this was conceivable, but only if we also assume that the speed of light propagating through the ether is unaffected by the changes in the density of the ether.

I will expound about the nature of this membrane including its formation in the chapter on gravity.

As we have observed, one need not be a rocket scientist to notice the vast difference between earth's atmosphere, which is below 100-km altitude, and outer space and the reason behind it. Hence, it is a real mystery why the scientific establishment ignores the various phenomena that occur only in outer space.

Experiments to verify the veracity of the new postulates

1) Michelson–Morley type experiment

If a similar experiment is conducted from an orbiting satellite at an altitude between 100 and 120 km, preferably at 110 km, from the earth's surface, where the medium is densest, it will definitely throw up positive results.

2) Experiment to find out if space is filled with an elastic medium

The substance existing between 100- and 120-km altitudes from the earth's surface should first be drawn into a sufficiently long pneumatic cylinder by pulling the piston outwards after which the inlet point has to be closed. Next, the piston should be pushed inwards. If space is really filled with an elastic substance, it will be compressed when the piston is pushed forward. Once the piston is released, the compressed medium will push back the piston just as in the case of compressed air. However, if space is truly empty, then the piston can be pushed inwards until the very end. Moreover, it will not revert once it is released.

3) Experiment to find out if the weightlessness experienced inside spacecrafts is really due to the absence of the elastic medium inside them

According to the new hypothesis, objects and astronauts experience weightlessness only while they are inside spacecrafts because they are shielded from coming into direct contact with the ethereal medium pervading space by the spacecrafts outer metallic shell. A simple experiment as described below can prove the veracity of this postulate.

A sounding rocket provided with a door should be fired to an altitude of about 115 km where the door should be opened to let in the space medium. If the newly posited theory is true, then all the unattached objects within the spacecraft will cease to experience weightlessness. Living in space will not only become less hazardous, but human being's aspirations of traveling to far away planets will become more feasible.

4) Experiment to prove that when a body enters space, it is still in possession of its entire weight

Although our earth itself is a classic example that proves that the weight of matter contained in a celestial body is always preserved and acts toward its center of mass, experimental results will strengthen the postulate further.

Items required

1) Two hemispheres with a combined diameter of about 250 mm made of a high-density material like lead with one of them having a cavity as illustrated in fig. 22 and a groove for the electrical wires.

2) A sensor is able to detect weight and transmit the signals through wires to an external device. It should be designed to fit into the cavity existing in one of the hemispheres and its height should be such that it allows the flat side of the two hemispheres to come within a few hundredths of a millimeter.

3) A digital processor that can process the signal and display the weight.

Fig. 22

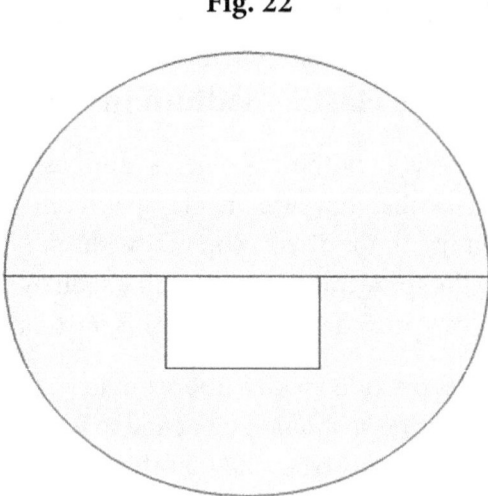

Experiment

The weights of the two hemispheres, along with the sensor, are first measured on the earth by using a single pan weighing machine after which they along with the other equipment should be taken to the ISS. After fixing the sensor inside the cavity with superglue, the two flat sides of the hemispheres should be pressed together with hands as illustrated in fig. 22. According to the new theory, even if the hands are now taken away, the two hemispheres will remain pressed together as a whole with a minute gap and also float giving the impression that they have lost weight. However, the external processor, which receives signals from the slightly compressed sensor, will now display the combined weight of the two hemispheres, which should be equal to the value obtained on the earth. This will not be possible on the earth's surface because the two halves will not behave as one but will fall apart.

If the result of this experiment is positive, which I am sure it will be, it will prove that just as bodies always possess mass, they are always in possession of their weight also. It is only due to the influence of external forces that the weights of objects appear to vary. It will also prove that a compressive force acts on all objects in space.

5) Experiment to prove that a gravitational field will develop around objects only when they are above 100 km from the earth's surface

A solid sphere of lead weighing about 10 tons or more should first be assembled from precast moulds on the earth's surface. If we place a small spherical object, like a marble, on it, we will notice that it will roll down and fall to the ground. Similarly, water or any other liquid will also fall to the ground if it is poured onto the lead sphere. The same lead sphere should be dismantled flown to the ISS and re-assembled outside it. A liquid other than water, which has a very low freezing point, should now be poured onto the lead sphere. If the new postulate is true, then the liquid will stick to and cover the entire surface of the sphere before freezing. Similarly, if a very tiny steel or glass ball is placed anywhere on the surface of the lead sphere, it should remain at the same place and not fall away. However, if the above postulate is wrong, then both the liquid and the ball will fall away into space. Sensitive instruments can also be used to detect the presence of a gravitational field around the lead sphere.

6) Experiment to prove that a compressive force from above is acting all around the earth's surface

Two weighing scales, two for balancing, should be assembled on either side of a sounding rocket as illustrated in fig. 23 onto which two round metal sheets having a diameter between 1 and 2 m are placed and secured with suitable glue. The metal sheets should preferably be of lead. Once the sounding rocket is fired and rises above an altitude of 80 km from the earth's surface, its speed should be suitably reduced so that it rises up with a steady velocity with zero acceleration. According to the new hypothesis, the weight of the sheet will gradually increase as the sounding rocket approaches the elastic membrane, which begins at 100-km altitude. This is because as the distance between the two decreases, the intensity of the mutual repulsion between them will increase and hence the maximum weight registered will be when the plates are just adjacent to the elastic membrane, that is, just below 100-km altitude from the earth's surface.

Fig. 23

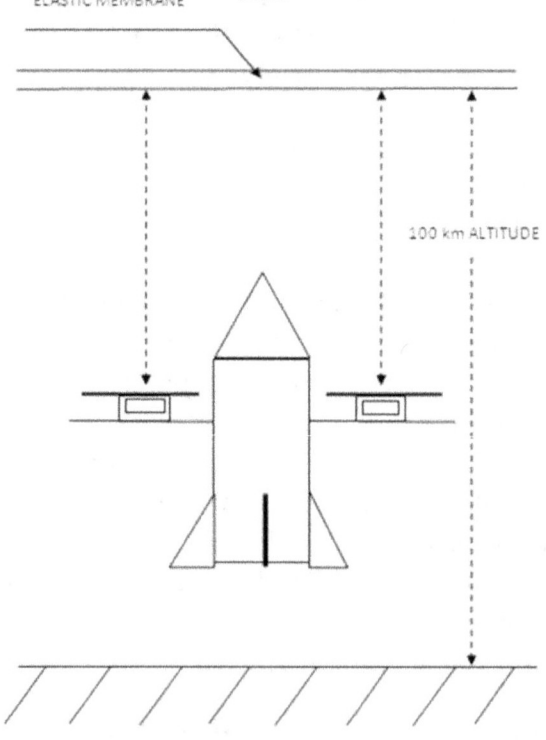

The result of this experiment will throw light on whether gravity is an attractive force emanating from the center of celestial bodies or a pressing force acting downwards from space.

Misconception Number 2

"The important thing is to not stop questioning; curiosity has its own reasons for existing."

Albert Einstein

Light needs a medium to travel

James Clerk Maxwell and several others believed that light required a medium to travel and this medium was the luminiferous ether. However, although most of the contemporary scientific community believes otherwise, there are a few, who still have faith in ether theories. This topic is important because the nondetection of this luminiferous ether led to the following presumptions.

a) Space is just an empty vacuum.
b) Speed of light is always constant and is independent of its source's as well as an observer's motion.

The following points prove that light does not need a medium to travel or propagate from one point to another.

1) We are now aware that although space is pervaded with an extremely low-density medium, this medium is absent in earth's atmosphere, which exists below 100-km altitude. The fact that light is still able to travel in our atmosphere proves beyond any doubt that no medium is required for the propagation of light.

2) It is also obvious that this medium in no way hinders the speed or passage of light because if it had, then we would not have been able to see the stars, which are billions of light years away from earth. In traveling

those vast distances, not only would light's speed have reduced, but its energy would have also been absorbed by the medium. From this, we can infer that the density of this medium is extremely low in spaces far away from celestial bodies as postulated by Sir Isaac Newton in his second edition of *Optics*. According to his later theory of gravity, the density of the medium around celestial bodies is much higher than in free space.

The various experiments conducted by Michelson and Morley repeatedly failed to detect ether drift because this medium is absent in the earth's atmosphere, that is, below 100-km altitude. Even if it had been present, it would have been difficult to detect, as it seems to travel along with the earth just as our atmospheric air and all the other atmospheric layers.

Misconception number 3

"A true seeker of truth, should at least once in his lifetime, doubt all things."

Rene Descartes

Matter is a wholly a condensate of electromagnetic energy

The eighteenth and nineteenth centuries are a golden era in the history of science as tremendous progress was achieved in all branches including physics.

However, the collective decision adopted during the same period of reducing all the mechanical properties of matter to that of electromagnetic just because the various experiments failed to detect ether is highly erroneous. This is because, as posited by Newton, Einstein, and various others, and from misconception number 1, we know that space is filled with an elastic medium of extremely low density.

Einstein beautifully explains the manner in which this misconception gradually crept into physics in his lecture. *"The development of the theory of electricity along the path opened up by Maxwell and Lorentz gave the development of our ideas concerning the ether quite a peculiar and unexpected turn. For Maxwell himself the ether indeed still had properties, which were purely mechanical, although of a much more complicated kind than the mechanical properties of tangible solid bodies. But neither Maxwell nor his followers succeeded in elaborating a mechanical model for the ether which might furnish a satisfactory mechanical interpretation of Maxwell's laws of the electromagnetic field. The laws were clear and simple, the mechanical interpretations clumsy and contradictory. Almost imperceptibly the theoretical physicists adapted themselves to a situation which, from the standpoint of their mechanical program, was very depressing. They were particularly influenced by the electro-dynamical investigations of Heinrich Hertz. For whereas they previously had required of a conclusive theory that it should content itself with the fundamental concepts which belong exclusively to mechanics (e.g. densities, velocities, deformations, stresses) they gradually accustomed themselves to admitting electric and magnetic force as fundamental concepts side by side with those of mechanics, without requiring a mechanical interpretation for them. Thus the purely mechanical view of nature was gradually abandoned. But this change led to a fundamental dualism which in the long-run was insupportable. A way of escape was now sought in the reverse direction, by reducing the principles of mechanics to those of electricity, and this especially as confidence in the strict validity of the equations of Newton's mechanics was shaken by the experiments with b-rays and rapid cathode rays.*

This dualism still confronts us in unextenuated form in the theory of Hertz, where matter appears not only as the bearer of velocities, kinetic energy, and mechanical pressures, but also as the bearer of electromagnetic fields."

The following points emphasize the fact that matter contained in objects cannot be just electromagnetic energy.

1) It is an undeniable fact that matter has dual characteristics. It is not only a bearer of electromagnetic fields but is also physical in nature and possesses mechanical properties like ductility, malleability, elasticity, and so on, which cannot be explained logically when we consider matter in terms of only electromagnetic energy that is concentrated in the incredibly tiny nuclei of atoms.

2) According to the equation $e = mc^2$, mass and energy are interconvertible. However, although an atom's nucleus can be split or even fused to release energy, it is impossible to convert electromagnetic energy into even a gram or milligram of the ordinary low-density matter existing in our bodies and all other objects present around us. If the equation was truly valid, then converting electromagnetic energy into normal matter should also have been a straightforward and easy task. Just visualize the FBBTS that our bodies are comprised of and the food we eat and

so on as electromagnetic energy, you will realize how unrealistic this assumption is. Moreover, it is a well-known fact that certain forms of electromagnetic energy are actually harmful to the ordinary matter present in our bodies.

Conclusion

These points clearly highlight the fact that energy in any form is somehow different from the ordinary low-density matter that we perceive all around us. We can easily convert mechanical energy into electrical and electrical into electromagnetic energy. This implies that we can also produce ordinary physical matter by using purely mechanical energy, which is highly illogical and an impossible task. Hence, there is definitely something more to matter that we need to understand properly, and that is how or what entity endows it with physical form?

From our discussion, it is now explicitly clear that the gradual agreement among physicists that matter is a condensate of electromagnetic energy only is not based on any solid scientific proof or theory. At best, it can be thought of as a hypothesis, which is yet to be proved, but has now ironically attained the status of being the sacrosanct truth.

Misconception number 4

"In questions of science, the authority of a thousand is not worth the humble reasoning of a single individual."

Galileo Galilee

The atom consists 99.9% empty space

Introduction

This misconception, which is a result of the first and third misconceptions, is probably one of the biggest misconceptions in the entire history of human civilization because this assumption is highly illogical as it fails to reflect physical reality. That matter is full of a low-density substance is not only a priori but even our logic, intuition as well as common sense tell us that it just cannot be 99.9% empty space. If someone tells us that an apple consists of only the seeds and a tiny portion of its skin present far away which is spinning so rapidly that it gives rise to the illusion that the apple along with all the pulp inside it is real, will we believe that person? Definitely not, we would instantly doubt her/his mental stability. Then, how is it that when the same thing is said about an atom, we believe it? We firmly believe that it is true simply because we have been taught so right from our childhood. Another factor that contributes to this huge misconception is psychological in nature. Most human beings are wired to always believe in authority or the official views and similarly conform to the opinions of the majority, as we feel uncomfortable otherwise.

There probably is not a single person in the world, who has attended school, who at one time or the other has not asked such questions; if matter consists of mostly empty space, why does it appears as being full and exists in a solid, liquid, and gaseous form? Why do not objects pass through each other? Why is it so difficult to cut metals like iron and steel even with a power saw? Why is not it possible to compress matter into a very dense state? How exactly does the addition of protons give rise to new elements? Why is the range of the strong nuclear so ridiculously short? The very fact that these and numerous other such questions do not have cogent answers proves that there is more to an atom than we presently know.

J.J. Thompson's atomic theory

In 1904, J.J. Thompson, a British physicist who had earlier discovered the electron, proposed a plum pudding model of the atom. The atom as a whole was compared to a round pudding, which was positively charged and the uniformly distributed plums represented the negatively charged electrons as depicted in fig. 24. As the positive charge of the pudding was cancelled out by the negative charge of the electrons, atoms as a whole remained neutrally charged.

Fig. 24

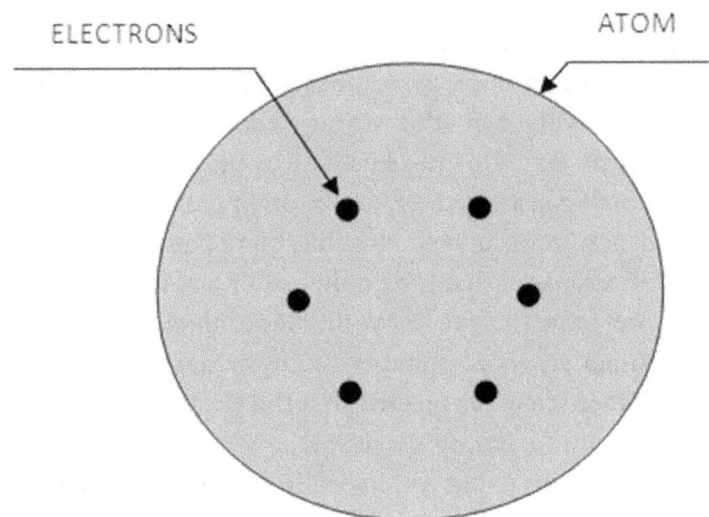

Ernest Rutherford's gold foil experiment

However, there was a paradigm shift regarding the structure of an atom 5 years later after Ernest Rutherford, along with his assistants Hans Geiger and Ernest Marsden, conducted the gold foil experiment. In this experiment, a thin sheet of gold foil surrounded on all sides a by zinc sulfide–coated wall was bombarded by alpha particles, which are actually positively charged helium nuclei. When the alpha particles struck the zinc sulfide–coated wall, a tiny flash would occur, clearly indicating the passage of the alpha particles through the foil. Rutherford, who until then believed in the plum pudding model, expected that all the alpha particles would pass through the gold foil and strike the zinc sulfide wall behind it. However, to his astonishment, he found that a few alpha particles were deflected and some even bounced back toward the source occasionally instead of just passing through the foil. The following were his words after observing this startling and unexpected result, *"it was like firing a cannon ball on a tissue paper and seeing it bounce right back at you on certain occasions."*

After this experiment, which he conducted several times with the help of his assistants to be 100% sure of the results, he concluded with the help of certain complicated mathematical calculations that the entire mass of an atom was concentrated in the dense nucleus existing in its center. The density of this nucleus is millions of times greater than the normal matter that we are familiar with or in other words its density is almost equal to the density of a neutron star. This nucleus is surrounded by a vast amount of empty space, which constituted 99.9999% of an atoms overall volume. (It never occurred to him or anyone else that there could be something else within the atom because at that time only matter was known to exist in the universe. It is only now that we know that other than matter, the universe should also contain an equal quantity of antimatter, dark matter, and so on.). Later on, other scientists posited that the electrons orbit the nucleus, similar to planets, in the empty space surrounding the nucleus. However, because this assumption could not logically explain why matter appeared as full or the manner in which the intermolecular forces worked, physicists came up with a new theory in which the electrons exist as an electron cloud, and this model of the atom more or less prevails even until today.

The discovery of the dense nucleus by Rutherford is definitely a momentous event in the history of physics. However, a part of the theory, which states that the atom consists of 99.9999% of empty space, is a grave misconception because it completely neglects the presence of the low-density elastic matter that we come across in our daily lives. **It is my humble opinion that Rutherford's discovery is much more momentous than thought earlier because it emphatically points to the existence of two different types of matter inside the atom.** One is the dense nucleus and the other is the low-density matter having elastic properties.

Reasons why atoms cannot be just empty space

The following reasons strongly affirm that atoms cannot be just empty but must be full of some unknown elastic material of extremely low density when compared to the nucleus, which gives fullness to objects and due to which matter is visible and appears as solid and liquid.

1) The first point of course is that this presumption does not reflect physical reality. Practical day-to-day experience tells us that objects are in fact full of a low-density substance and not empty. Hence, how can this miniscule amount of super dense matter contained in the nucleus alone, its density is equal to the density of matter present in neutron stars, appear as the extremely low density and elastic matter like that present in the (FBBTS) of our bodies? What is it that we eat, drink, and excrete? It definitely cannot be the atom's nuclei or electrons alone because there combined volume is almost 100,000–300,000 times lesser than the actual volume of an atom. We often read in articles that if all the matter contained in the atoms of the entire human civilization were to be brought together, it would fit into a small teaspoon.

   The universe consists of more than 99% of supposedly empty space and stars, solar systems, galaxies and black holes, and so on make up the remaining. Because all these heavenly bodies comprise of only atoms, which are again supposed to contain 99.9% empty space, how can this ridiculously tiny amount of super dense matter, contained in the nuclei, alone make up the entire universe and give rise to the force of gravity that spans across billions of light years?

2) The alpha particles used in Rutherford's experiment, in atomic terms, are massive in size when compared to the extremely tiny light particles called photons and possessed a speed of almost 19,000 km/sec. If these massive particles could easily pass through the 2-micron thick gold foil used in the experiment, then logically even light should be able to easily pass through the foil because the photons are miniscule in size when compared to the alpha particles. However, in reality, we are all aware that light will not pass through a gold foil or for that matter any other metallic foil, however thin it is. In fact, most of the light is reflected back by the foil.

How then could the massive alpha particles penetrate the gold foil in the experiment? They have penetrated because of the huge mass and high velocity possessed by them and not because the atom's nucleus is surrounded by empty space. It is my contention that if a similar experiment were to be conducted today where the speed of the alpha particles is somehow reduced sufficiently, then none of the alpha particles would be able to pass through the gold foil but would instead bounce backwards just like the light rays.

Fig. 25

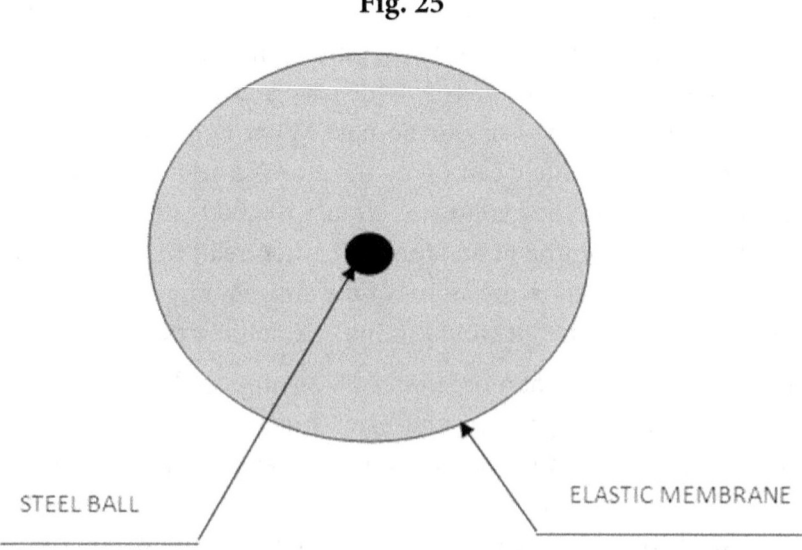

STEEL BALL   ELASTIC MEMBRANE

A simple thought experiment will demonstrate what is being implied. Let us visualize a steel ball attached to a thin elastic membrane as illustrated in fig. 25, which represents an atom. If we throw a small steel ball with our hands against the elastic membrane, it will bounce back toward us because of insufficient velocity. However, if we fire the same ball from a pistol, it will effortlessly tear through the membrane. It will be deflected or might completely rebound if it happens to strike the steel ball depending on its point of contact with it. Although, an X-ray reveals only the dense bone structures in our bodies, we know that this structure is surrounded by flesh and blood. The gold foil test is almost similar to an X-ray. The nature of the experiment is such that it can detect only highly dense substances.

3) Because the nuclei of atoms are a condensate of electromagnetic energy, they are highly explosive in nature. In fact, we will not be wrong if we refer to them as tiny nuclear bombs. The only reason they do not explode is because the mysterious strong nuclear force, one of the four fundamental forces of nature, is holding them together. A strange and inexplicable aspect of the strong nuclear force is that although it is supposed to be millions of times stronger than gravity, its range is extremely short. In fact, it never extends beyond the nuclei of atoms. This is a puzzling phenomenon because in nature it is generally found that the range of any force increases with increase in its strength and vice versa. For example, a powerful magnet will be able to attract or repel from a greater distance than a weaker one and so on. Moreover, the range of all known forces do not cease abruptly as is in the case of the strong nuclear force. Normally, the strength of a force will gradually decrease as the distance from the source increases and vice versa in accordance with a particular ratio, for example the inverse square law and so on.

Einstein, once in connection with quantum physics, has commented as follows. *"The theory says a lot, but does not really bring us any closer to the secret of the old one. I at any rate am convinced that He (meaning God) does not play dice."* Similarly, we will not be wrong if we state that nature does not function in an ad hoc and whimsical manner meaning that all forces should function and behave in the same way.

4) Another strange aspect associated with an atoms nucleus is that two forces seem to emerge simultaneously from within it; one repulsive and

the other attractive. The positively charged electrostatic force causes two nuclei to strongly repel one another. However, when the two nuclei are brought extremely close to each other, then the highly attractive strong nuclear force takes over and binds the nuclei together. Because all known forces must come under one of these three categories, namely electromagnetic, electrostatic, and mechanical, let us try to find out to which of these three the strong nuclear force belongs. It cannot be an electromagnetic force as it is present in the atoms of all known elements. It cannot also be an electrostatic force because for one, its range is too short, and second, if both a positive and a negative electrostatic force arise from the same source (i.e., the nucleus), then the weaker of the two forces, the Coulomb force, should be neutralized which is not so in this case. This leaves us with only one option, which is that the strong nuclear force is a mechanical force. I will expound about the mechanism behind this force later on in this chapter.

5) We know that with every addition of a proton, starting from the hydrogen atom, a heavier element is formed with entirely different physical, mechanical, chemical, and electromagnetic properties. We can understand an increase in mass, but how exactly does this internal transformation takes place? The current atomic theory does not delve into these matters. Although we know that it is the number of protons present in an atom's nucleus that decides which element is formed, the internal process by which one element transforms into an entirely different element by the addition or subtraction of protons is not clear. Currently, there is no convincing explanation for this in physics. A comment posted by a student, regarding this particular query, in the Internet is as follows. *"I asked this question to two physics teachers and three chemistry teachers but did not receive a positive reply. Instead each one told me to ask the other."*

It is an established fact that the basic nature and composition of the protons, neutrons, and electrons remain the same irrespective of whichever element they are a part of. For example, every proton present in the nucleus of a gold atom is an exact replica of the proton present in a hydrogen atom. The same applies for neutrons and electrons also. Only their quantity will increase as the atomic number of an element increases and vice versa. Hence, it is impossible, logically as well as scientifically, to explain how the same basic three ingredients can give rise to the various elements.

6) The nuclei containing protons and neutrons always remain in the center and are surrounded by the electrons or electron cloud and hence everything that we see and feel must always be the electrons? What causes these electrons to appear as different elements and substances? When a combustible substance burns giving off energy in the form of heat and light, how and what exactly is burning? It definitely cannot be the protons, neutrons, or electrons because as mass is always conserved, the combined mass of the by-products released and left over after burning will always be equal to the initial mass. Moreover, these particles by themselves are not flammable.

However, one thing is certain and that is, the ordinary matter found in all objects is definitely not the negatively charged electrons or electron cloud as is currently theorized because of the following reasons.

   a) Individual electrons have a mass of only 1/1,836 of the mass of a proton, which is so negligible that it is not even considered while calculating the atomic weight of atoms and hence they alone cannot give fullness to objects. If an atom's nucleus is compared to be the size of a football (22 cm in diameter), then the nearest electrons, which are 1,836 times smaller than the football, would even then have a diameter of only about 0.12 mm approximately and would be orbiting at a distance of almost 0.75 km away from the football. Forget 0.75 km, take a football in one hand and a few grains of sand or mustard seeds in the other and stretch it out. Now try to visualize these mustard seeds spinning around the football to give rise to a solid three-dimensional sphere and you will realize the enormity and impossibility of the task.

   b) If it is true that objects appear full and exist as solids and liquids only because of the electrons, then all the objects that we see around us including our own bodies, the food and beverages that we consume, and so on are not real but mere illusions created by the whirring electrons. Just visualizing this thought is so unbelievable and illogical.

7) An atom is defined as the smallest indivisible part of an element that is in full possession of all properties. Hence, it goes without saying that although it is so unimaginably small, it must still possess appearance including color and all other physical, mechanical, chemical, and electromagnetic properties of whichever element it belongs to. In other words, it should contain at least some quantity of ordinary low-density elastic matter, which make up our bodies and all other objects by virtue of which when trillions of them are bunched together, they appear as the various elements.

For example, even though we cannot see individual particles of talcum powder, which measure about 10 microns with our naked eye, when millions of them are brought together, we can see them collectively as a white powder. Similarly, gases, which are invisible and comprise of individual atoms and molecules will collectively become visible as liquids when they are sufficiently compressed and cooled. This means that when we look at an objects surface, even when only with our naked eyes, we are actually observing trillions of atoms sticking to each other. Hence, logically speaking, individual atoms must be a microscopic replica of that which we are observing.

8) Now let us discuss a little bit about the intra and intermolecular forces, which are supposed to be electrical in nature and bond atoms and molecules to each other. There are numerous dichotomies in the present theories as the following question and answers will reveal.

Question: What causes neutral atoms to stick to each other?

Answer: It may sound strange that although the atoms are surrounded by the negatively charged electrons, they do not repel each other. Instead, the electrons in one atom are attracted to the protons in another atom, an effect that is called as the residual electromagnetic force, due to which they stick together.

Question: If atoms are mostly empty space, why don't objects pass through one another?

Answer: They do not pass through other things because they are levitating on an electrostatic field. When you sit on a chair, you are not really touching it. Because a shell of electrons that are negatively charged surrounds the nuclei of atoms, they repel each other even though the atoms as a whole are neutrally charged.

Question: In that case, why does the entire world not blow away from itself?

Answer: Most atoms' electron shells are not full. When two atoms come together and have empty spaces in their electron shells, they will share electrons to fill the empty spaces in both their shells. Electrons tend to go back and forth between atoms very fast.

Question: Do molecules also stick together in the same manner as the atoms?

Atoms are bonded together by the intramolecular forces, which are electrostatic in nature. The intermolecular forces bond molecules together where the molecules behave as dipole magnets. One end of each molecule induces a temporary dipole in the adjacent molecule due to which they are attracted to each other as illustrated in fig. 26.

Fig. 26

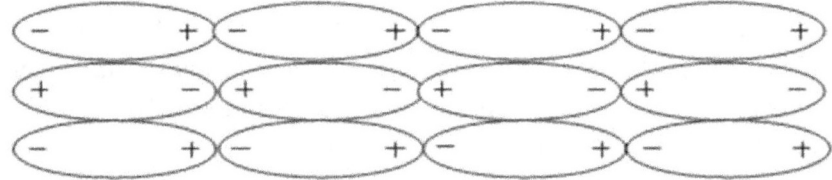

Although the electrons are in rapid motion (electrons are supposed to be moving at 2,500 km/sec), the molecules are still attracted to each other because the process of temporary inducement also occurs very rapidly, instantaneously, and in perfect synchronization between the trillions and trillions of atoms present in matter. Even if we consider this explanation as logical in the case of solids, how will molecules align themselves so perfectly in liquids where their positions are constantly changing? Moreover, according to quantum mechanics, electrons are not point particles but exist as waves or electron clouds.

The various answers clearly highlight the dichotomies existing in the various theories that try to explain how the intra and intermolecular forces function. However, the biggest dichotomy of all is that all these atoms and molecules are electrically neutral, which means that the negative charge of the electron has been neutralized by the positive charge of the proton. When such is the case, it is not wrong to assume that they are still in possession of their electrical charges? Moreover, if this assumption is true, then what entity within the atom has become neutral?

The very fact that the intensity of the intermolecular force is mostly dependent on the temperature of a body; it increases with decrease in temperature and vice versa, and to a lesser extent on its composition does not support the current theory. Electromagnetic and electrostatic forces, in reality, are not so highly sensitive to heat.

9) The elastic collisions between the gas atoms and molecules clearly illustrate the fact that they behave like individual rubber balls. This would not have been possible if atoms really contained mostly empty space and a few rapidly spinning electrons.

10) If atoms truly contain only a dense nucleus surrounded by a few electrons in the distance separated by empty space, then it should have been easy to achieve nuclear fusion, at least on an experimental basis, by transmutation where the protons are captured by the strong nuclear force when they strike the nucleus. A continuous stream of protons could be fired, at the required speed, at different substances for days on end after which they could be tested to see if transmutation has occurred. The very fact that transmutation cannot be achieved easily implies that the nuclei are not bare but are surrounded by an unknown substance.

11) Today optical microscopes are available that can magnify up to 1,500 times. Electron scanning microscopes have such a high magnification and resolution that scientists can now observe even a single atom. Hence, we should have been able to perceive the emptiness of atoms or matter under these microscopes. However, even under these microscopes, matter contained in objects does not appear as empty space.

12) As the strength of mechanical forces can far exceed the strength of electromagnetic and electrostatic forces, which is supposed to be preventing atoms and hence matter from collapsing, it should have been pretty easy to compress matter, using hydraulic presses that can generate pressures exceeding thousands of tons per square centimeter, into a super dense state which however is impossible.

13) According to the current theory, there are seven energy levels or electron shells in an atom and electrons can be present only in these shells and cannot exist in the empty space in between them. As far as the theory goes, it is fine. However, there is no explanation as to how or what forces give rise to these electron shells. It is impossible for simple electrostatic or electromagnetic forces to give rise to and sustain such a complex structure.

14) The very thought that all matter is a soup of only protons, neutrons, and electrons, which are lifeless entities, is incompatible with biology, DNA, the intricacies of the human body including the brain, intelligence, and life itself.

The various points and thought experiments clearly highlight the deficiency in Rutherford's theory as far as the empty space in atoms is concerned. Other than the dense nucleus and electrons, which represent energy, an atom must contain a low-density gel-like substance, whose collective density or hardness is measured by using the Rockwell and Vickers scales and which we normally refer to as matter. This is why matter exists as solids, liquids, and gases and is both a bearer of electromagnetic fields as well as mechanical qualities.

Rutherford and others believed in the new theory because at that time only matter was known to exist.

However, today, thanks to Paul Dirac and the tremendous advancement in physics, we know that an equal quantity of antimatter also must exist. Why this seems to be missing from our universe is one of the biggest mysteries in the world of physics.

We can still compare the atom to the plum pudding model proposed by J.J. Thomson, which reflects physical reality and makes more sense. Just as in the plum pudding model of the atom proposed by him, it is more logical to presume that the electrons are embedded in this unknown material in a definite order or pattern than to visualize them as speedily rotating particles or as electron clouds. For example, if we visualize 80 electrons rotating at high speed, in the various energy shells, without colliding with one another in the tiny space existing within a gold atom, we will understand the enormity and complexity of the task.

Keeping these various dichotomies and discrepancies in mind, I am proposing a new hypothesis regarding the structure and functioning of an atom, in which there is absolutely no room for ambiguity.

However, before we proceed with the new hypothesis, it would be of immense help if we could get even a faint idea of the nature of the low-density substance present in atoms. For this, we have to first find out if the supposedly empty space in atoms is the same as that present in outer space. If it by chance happens to be the same, then we will know that the low-density substance present in atoms is a condensate of the substance pervading outer space.

Is the empty space present in atoms the same as that present in outer space?

Although this task initially seems daunting, we can easily find out by the following reasoning.

On the earth's surface, when the mass of an object of a same substance increases, the volume of the object will increase proportionately, whereas its density will remain constant. But in space, as the mass of huge bodies like our earth and planets increase, their volume would not increase proportionately, but instead their average densities will increase. Although density in both the cases means mass per unit volume and is expressed as g/cm^3 in the metric system, there is a subtle difference between them, which we should be aware of.

When we talk of the density of elements on the earth, we are actually referring to the density of an atom's nucleus that depends on the number of nucleons it contains. For example, although there is just a little difference between the atomic radii of aluminum and gold atoms, the nucleus of a gold atom contains 276 nucleons against the 40 present in the nucleus of an aluminum atom. This means that a gold atom is almost seven times heavier than an aluminum atom because of which the volume of a unit mass of gold will always be lesser than the volume of a unit mass of aluminum as illustrated in fig. 27.

Fig. 27

X gm AU = 40 atoms   X gm AL = 280 atoms

When we refer to the average density of planets, we are actually referring to the average atomic radii of the atoms that constitute the overall mass in a planet. As we move closer to the center of mass of a planet, say our earth, the density of matter keeps increasing because the pressure on matter is also increasing. Density here means that the distance between the various atoms' nuclei is reducing. The density of matter in the core of the earth will be the highest because it is subjected to a pressure of almost three million atmospheres. Because the nucleons present in the atoms of the various elements that constitute our earth always remain constant, there is only one change that can affect the density of matter and that is the atomic radii of the atoms. This means that the atomic radii of the atoms present in the earth's core will be the least (matter here will have maximum density) and as we move toward the surface, atomic radii will keep increasing until at the surface of the earth the atoms will have maximum atomic radii. (Matter here will have the lowest density.)

Now let us find out what actually happens when the mass of a planet increases.

The earth, like all other heavenly bodies, has a particular mass, average density, weight, and volume. Instead of going into actual figures, let us denote the above-mentioned physical properties of our earth with symbols as noted below:

Mass = m

Average density = d

Volume = v

Weight = w

Let us assume that another body, which is a replica of earth in all aspects, is merged with our earth. The resulting changes that occur in the new body will be as follows.

Mass = 2 m

Average density = d + an unknown factor

Volume = 2 v − an unknown factor

Weight = w + an unknown factor

In other words, the mass of the newly formed body has exactly doubled, whereas its average density and weight have increased more than what should have been. Because of this, volume that should have also doubled is actually much lesser. This is similar to what happens to atoms when they fuse together to form heavier elements. Their volume does not increase in proportion to the combined volume of atoms in a free state.

From this, we can infer that when the mass of heavenly bodies increase, the atomic radii or the size of the atoms that constitute it will become increasingly smaller because a part of the empty space associated with them gets displaced to outer space.

To find out if this is really true, let us consider another example.

It is stated in physics books that if all the empty space present in the atoms of our earth were removed, the diameter of earth would be about 19 mm only. This is actually known as the Schwarzschild radius. The Schwarzschild radius is a radius of a sphere such that, if all the mass of an object were to be compressed within that sphere, the escape velocity from the surface of the sphere would be equal the speed of light. According to information available on the Internet, the Schwarzschild radius of our sun is about 3 km only.

This is not just a figment of someone's imagination or a mere theory put forward by scientists and astrophysicists because they have actually discovered such dense bodies existing throughout the universe with the help of sophisticated instruments. Some of the dense structures whose diameters could be in the range of only 3–10 km are known as pulsars, neutron stars, and stellar black holes. The original size of these dense bodies would have been much more than our sun whose diameter is about 1.392 million km. Here, we are again confronted with the question, what has become of the empty space that was part of the massive body, before it collapsed to its present dense state? The answer is now quiet obvious. The displaced empty space has merged and become a part of the vast outer space, proving beyond any doubt, that the empty space present in atoms is the same as that present in outer space.

We can now safely infer that, irrespective of whether a gravitational force exists or not, the weight of an object with constant mass will increase only if its density increases and vice versa. Moreover, for density to increase, the atomic radii of the atoms contained in the objects should decrease.

If we now think of the composition of our bodies, the actual matter in our bodies is a million times smaller than a grain of sand and the remaining empty space is the same as that present in outer space. From this, we can also infer that the second matter present in outer space is also present in the empty space of atoms of our bodies in a condensed form just as the nucleus of an atom is a condensate of electromagnetic energy.

New Atomic Hypothesis

Before ether theories were banned, most scientists, physicists, and philosophers believed that ordinary matter must contain ether. Rene Descartes went so far as to say that matter is purely made out of ether, and Newton postulated that it was denser in gross bodies than in free space. Even Nikola Tesla hypothesized that matter was a condensed form of the ether pervading space. The belief that matter existed only in the nuclei of atoms began after Rutherford discovered that atoms have a dense nucleus.

The new hypothesis, which is a result of combining the earlier belief and Rutherford's atomic theory, is based on following undeniable fact.

As matter possesses both mechanical and electromagnetic properties, it has to be constituted by two entities. It is a proven fact that the nucleus of an atom is a condensate of electromagnetic energy that fills the whole of space. Similarly, the low-density substance having elastic properties, which surrounds the nucleus, must also be a condensate of some rarefied substance that also pervades the continuum of space but does not possess electric charge. That is why although energy and matter are thought to be interconvertible in reality, energy in any form can never be converted into the normal low-density matter present in all objects and which is also why the equation $e = mc^2$ is unidirectional.

Hence, according to the new hypothesis, an atoms nucleus represents energy, whereas the low-density matter surrounding it represents physical form. It alone endows matter with fullness. Let us refer to this new substance as second matter (SM).

As the mere presence of these two entities will not explain why an atom's nucleus stays in a compressed state, we have to presume that they must also be intensely repelling one another just as the earthly matter and the elastic membrane existing between 100 and 120 km from the earth surface.

Fig. 28 illustrates the structure of an atom according to the new hypothesis.

Fig. 28

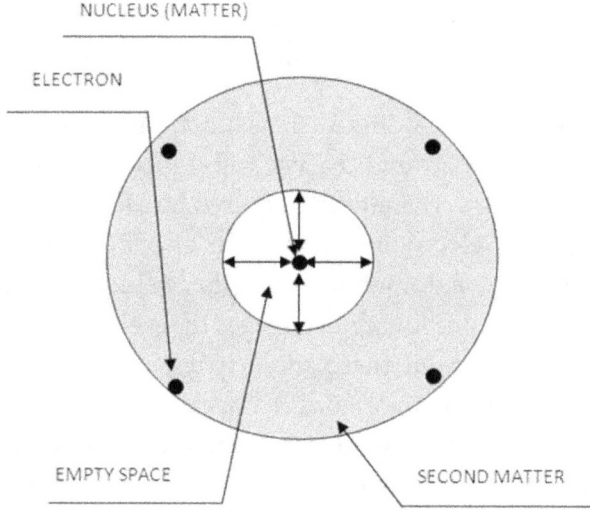

According to the new hypothesis, the dense positively charged nucleus now represents the plum and the unknown neutrally charged material, which we have named as the second matter, would represent the pudding. The intense repulsion between the two will give rise to two things. A massive compressive force is exerted onto the nucleus, which may be said to represent the strong nuclear force, due to which the highly repulsive forces between the nucleons are overcome and the nucleus remains in a dense and stable state. This is why the influence of the strong nuclear force, which is supposed to be millions of times stronger than gravity, never extends beyond an atom's nucleus. At the same time, the intense repulsion causes the highly elastic and tenacious material to stretch outwards due to which an area of empty space arises between the two. The atom will now appear as a solid sphere although a tiny portion of its inside is empty space (vacuum) as illustrated in fig. 28.

As everything existing in nature must have a specific purpose and function, let us try to find out the functioning of the newly hypothesized atom.

Functions of an atom's nucleus

a) The main function of the nucleus of an atom is to act as a tiny storehouse of extremely concentrated energy that drives the entire universe.

To truly understand how an atom's nucleus functions as an energy reservoir, let us observe the world around us. We will notice that the driving force of the universe is energy. The energy required by electrical and electronic goods and gadgets is electricity, which can also be stored in batteries. The energy required by automobiles and other vehicles is stored in fossil fuels like petrol and so on. Living beings get their energy from the energy stored in food. However, all these three types of energy are ultimately derived from the sun or in other words from the nuclei of atoms present in the sun. Now let us visualize a world where only energy in the above forms was to exist and the physical entities (e.g., the vehicles and living beings) that use this energy were to vanish. In such a case, energy by itself would be of no use and would serve no useful purpose; a physical entity is always required, which consumes this energy or onto which the energy acts upon or is stored in.

Another interesting aspect of all these forms of energy is that they will rapidly deplete if used. Hence, wouldn't it be wonderful if we could store thousands of times more energy in a small battery than is possible today? Similarly, wouldn't it be great if we could compress thousands of liters of fossil fuels to fit into the volume of 1 liter and a full year's food in a tiny box?

Although we humans are incapable of achieving this feat, nature has perfected this art to perfection and the result is the nucleus of atoms. It is an established fact that the nucleus is a highly concentrated form of electromagnetic energy and hence we can compare each nucleus to a tiny battery, which has a mind-boggling amount of energy stored in it. This energy that is locked up in an atom's nucleus can be released artificially by a man-made process called nuclear fission in which the nuclei of heavy elements like uranium are split by bombarding them with neutrons. However, nature releases this energy naturally in two ways both of which are triggered when mass increases beyond a certain limit. Heavy elements having an atomic number of 100 and above (each nucleus of such elements contain more than 170 nucleons) experience radioactivity due to which energy is released in the form of heat, along with the emission of alpha, beta, and gamma particles. Similarly, when the mass of celestial bodies, irrespective of their composition, equal or exceed our sun's mass, they will also transform into a nuclear furnace gradually releasing the locked up energy over billions of years.

b) Protons present in an atom's nucleus emit a positive charge, which is exactly equal to the negative charge of an electron because of which an atom as a whole will remain electrically neutral. The positive charge radiated by it also has a vital role in deciding the elemental nature of atoms.

Functions of the second matter surrounding an atom's nucleus

a) More than 99% of an atom is filled with this second matter because of which physical objects appear as being full. It endows objects with all their physical properties by transforming into the various elementary substances. It is solely responsible for binding atoms and molecules together.

b) It acts as an housing (a balloon with an extremely thick membrane) in which the highly explosive electromagnetic energy is stored in a dense and solid state called nucleus. The intense and mutual repulsion between matter contained in the nucleus and the second matter surrounding it gives rise to this massive compressive force. This once again explains why the range of the strong nuclear force never extends beyond an atom's nucleus.

Internal mechanism by which one element transforms into another element

This subject is completely absent from the current atomic theory.

The process could be as follows. Although the protons always remain inside the nucleus, each proton radiates a definite amount of positively charged energy, which is exactly equal to the negative charge of an electron, in the form of photons. This energy is absorbed by the neutrally charged second matter surrounding the nucleus and transforms it into different elements. For example, a single proton and an electron along with the second matter form a hydrogen atom and the positive energy radiated by the lone proton is absorbed by the second matter, which then becomes positively charged. However, as this charge is equal to the negative charge of the electron embedded in it, the atom as a whole, that is, the second matter, will remain neutral. When four hydrogen atoms fuse together, the following transformations take place. Two protons will fuse with two electrons to form two neutrons. In a similar manner, the second matter present in the individual hydrogen atoms will also merge into a single entity thereby increasing its density. As the nucleus now has two protons, the energy radiated exactly doubles and is absorbed by the denser second matter thereby forming the neutrally charged helium atom in the process. That is why when atoms fuse together to form heavier elements, their volume does not increase proportionately when compared to the combined volume of atoms in a free state, but instead their density increases. This process repeats itself as more and more protons and neutrons are added to the nucleus. The density of the second matter is also constantly increasing due to the merging of the second matter surrounding the additional nucleons resulting in the formation of heavier elements with entirely different electrical, chemical, and mechanical properties.

In this beautiful and elegant manner, interaction between energy radiated by the protons in the nucleus, which is a condensate of electromagnetic energy and the second matter surrounding it, which is a condensate of the ethereal medium pervading space, nature has enabled the creation of the wonderful, beautiful, and mind-boggling physical substances that make up the universe and of which we are an integral part.

Half-life of elements

Most elements along with their isotopes, including even very stable elements, have half-lives due to which they gradually degenerate into elements with smaller atomic numbers. It is easy to understand the degeneration of radioactive elements, but why do even stable elements degenerate albeit very slowly. According to the new hypothesis, the second matter surrounding an atoms nucleus needs a continuous supply of energy in the form of photons for it to retain its elemental form. The protons present in the nucleus emit the required energy because of which even though there is no noticeable radiation from stable elements their net energy level is gradually decreasing over long periods.

Why elements with higher atomic number become radioactive?

As more and more atoms merge together to form heavier elements, the intensity of repulsion between the nucleus and the second matter surrounding it will also be increasing proportionately. Because of this, the distance between the outer surface of the nucleus and the inner surface of the second matter will also increase gradually as illustrated in fig. 29, thereby decreasing the intensity of the compressive force exerted onto the nucleus.

Fig. 29

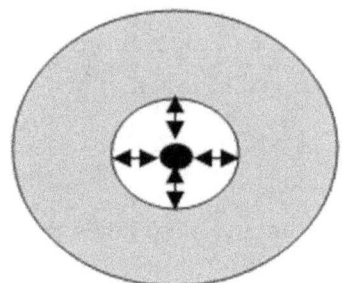 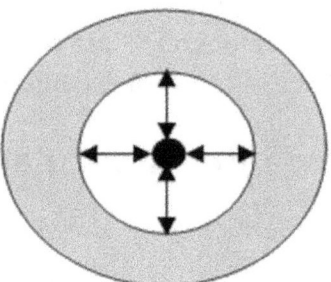

However, the mutual repulsion between the nucleons within the nucleus will increase unabatedly and at a particular point will equal or exceed the compressive forces exerted on it due to which a few elementary particles present in the nucleus gain sufficient kinetic energy to tear through the surrounding material (SM) and escape outside the atom. This element has now become radioactive.

This clearly highlights the fact that the strong nuclear and the weak force are not independent forces as such but are an indirect effect arising from the intense repulsion between the matter contained in the nucleus of an atom and the second matter surrounding it. In other words, they are mechanical forces.

According to the new hypothesis, the nuclei of all atoms beginning from the hydrogen are bound together in the same manner, and hence, all atoms can be split and in the process, every one of them will release energy. As mentioned, all atoms are similar to gas-filled balloons. The balloon membrane (second matter) in the hydrogen atom is the thickest when compared to the radius of its nucleus and thinnest in the atoms of uranium and other heavy metals, which are like balloons in which the gas pressure is so high that they are on the verge of exploding. Hence, it is easier to split atoms of heavier elements than atoms of lighter elements. This is why the proton is the most stable particle; it has the highest half-life when compared to other fundamental particles.

Thermonuclear weapons release excessively greater amounts of energy than purely fission weapons probably because the massive heat and compressive force generated by the primary fission explosion completely destroys the second matter present in the deuterium atoms due to which entire protons burst into pure energy. Of course, not all the protons present in the fuel comprising deuterium might be destroyed in this manner.

How intermolecular forces work?

The intermolecular force bonds individual atoms and molecules to each other to form the various elements and compounds. The strength of this bond is greatest in solids and lower in liquids. As the temperature of a solid body is increased by heating, the strength of this bond decreases proportionately, and at a particular temperature, the solid body transforms into a liquid. This is known as the melting point of that particular material. If this liquid is further heated, it will vaporize beyond a certain temperature and this is called as the vaporizing point. The exact reverse will occur when gases and liquids are cooled.

Based on these observations, we can safely infer that the intermolecular force is not electromagnetic in nature but arises from the adhesive nature of the second matter surrounding an atom's nucleus. This substance having adhesive properties hardens with decrease in temperature and vice versa.

That is why when two perfectly flat surfaces of any metallic plate or block are pressed together, they neither attract nor repel each other, whereas if the bonding forces had been electromagnetic in nature, then some sort of interaction would definitely have occurred. The only method by which they can be transformed into a single homogeneous plate is by melting them together and then allowing it to cool.

Comparison between current (CAT) and the new atomic hypothesis (NAH)

a) CAT - An atom consists of only an unimaginably tiny nucleus with electrons orbiting at a vast distance away from it. This does not conform to reality and does not satisfactorily explain as to what all visible and physical matter is made of.

NAH - The empty space surrounding the nucleus of an atom is filled with neutrally charged second matter, which gives fullness to atoms and in which the electrons are in constant motion in their respective energy shells. The matter contained in the nucleus is exclusively a source of energy, whereas the second matter gives physical form to objects. In conjunction with the energy radiated by the protons, the different elements are formed. Hence, it is the second matter that is perceived everywhere around us and not the protons, neutrons, or electrons.

b) CAT - Attractive forces present within the nucleus itself hold together the dense nucleus. It does not specify whether the strong nuclear force is mechanical or electromagnetic in nature. This assumption also does not logically explain why the range of this extremely powerful force is always confined or effective only within the nucleus. Moreover, independent attractive forces are alien to nature and only exist in theory. There is no experimental proof to validate this theory.

NAH - The mutual and intense repulsion between the dense nucleus and the second matter gives rise to a massive compressive force that acts on the nucleus from outside and holds it together. This presumption not only logically explains why the range of the strong nuclear force never extends beyond the nucleus but also clearly specifies that it is a mechanical force.

c) CAT - Does not explain the internal mechanism by which new elements are formed with the addition of protons.

NAH - Explains in a logical manner how new elements are formed with the addition of protons.

d) CAT - Intermolecular forces are purely electrostatic or electromagnetic in nature. This does not explain why its strength decreases with increase in temperature and vice versa.

NAH - Intermolecular forces are a result of the adhesive nature of the second matter present in atoms. An appropriate example is soap bubbles sticking to one another. This substance is sticky in nature and hardens with decrease in temperature and vice versa. However, the adhesive properties of the second matter of a particular element depend on the number of protons present in the nuclei of atoms of that element. Electrons of course play a vital role in the chemistry of elements and compounds and their molecular structures.

Conclusion

If you have read until here, I can imagine the expression of bewilderment and intense disbelief on your faces. Although the new hypothesis is more logical than the present theory, it may sound preposterous to most of you.

However, if you read the following quote by Dresden James, you will begin to understand why you feel like that.

"A truth's initial commotion is directly proportional to how deeply the lie was believed. When a well-packaged web of lies (knowingly or unknowingly) *has been gradually sold to the masses over generations, the truth will seem utterly preposterous and its speaker, a raving lunatic."*

Here I am not suggesting that Rutherford lied when he stated that an atom consists of more than 99% empty space. He was compelled to say and anybody else in his place would have also said the same because of two things. Firstly, the mechanical nature or characteristics of matter had already been reduced to that of electromagnetism long before he conducted this experiment. Secondly, at that point in time, unlike now, only matter was known to exist.

Although we humans are the most intelligent species existing on earth, we are highly susceptible and can easily be influenced by people in authority because our brains are wired that way. People who have watched programs such as *Brain Games* and *Our Beeped up Brains* on the TV will know what I am inferring to. The atom consists of 99.99% of empty space is a classic example that stands testimony to the validity of the following quote by Einstein. *"Unthinking respect for authority is the greatest enemy of truth."* We believe that all the physical things that we are actually observing, feeling, eating, drinking, and tasting do not really exist. On the contrary, we believe that visible matter is truly, made up of only protons, neutrons, and electrons, which are entities that we cannot directly observe or feel only because the textbooks and the people in authority are saying so.

What is indeed true is that both are real and both exist. In fact, without the one, the other also cannot exist.

I now request all my esteemed readers to introspect in a pragmatic and unbiased manner, apply the principle of Occam's razor and pass your judgment as to which of the two theories is more logical and believable.

Experiment to test the veracity of the new atomic theory

Even though a hydrogen atom possesses only one electron, the gas itself will transform into a liquid when it is compressed and cooled with the required pressure and temperature. According to current theories, the electrons are the ones that endow volume to atoms. According to this presumption, if a hydrogen atom is stripped of its electron, its atomic radius will immediately decrease by 100,000 times as it is now left with only a proton. Hence, even if zillions and zillions of them were to be collected and compressed as well as cooled appropriately, they should never condense into a liquid state. However, according to the newly posited hypothesis, the protons will still transform into a liquid and its volume will be equal to the volume that would have resulted without the electrons being stripped from the atoms.

I do not know the practicality of this experiment, but given the highly advanced technology that is presently available, it should be possible if the experimental volume is restricted to producing only a few milliliters of ionized liquid hydrogen. This experiment can be conducted even with alpha particles.

This experiment is of fundamental importance because negative results will further strengthen the prevailing theory, and scientists and physicists can rest assured that they are on the right track. However, if the new postulate turns out to be true, then they must decide upon the next course of action.

5

Misconception number 5

> *"The exact opposite of what is generally believed is most often the truth."*
>
> Jean de la Bruyere

Gravity is an independent attractive force

Introduction

This is the fifth biggest misconception because according to the theory, every object attracts every other object in the world. This is impossible because gravity is the weakest of the four fundamental forces of nature. In fact, it is so weak that a small magnet is able to lift up a ferrous object line a nail or pin acting against the entire gravitational pull of the earth. That is why it has not been detected even until today. Every one of us thinks Sir Isaac Newton perpetuated this myth. However, as we have discussed earlier, he himself never really believed that this was possible and hence came up with alternate theories, which however never gained acceptance.

Hence, Albert Einstein first broke the myth of gravity by suggesting that it was not an independent attractive force but a curvature of space-time. Although his theory of gravity is supposed to have superseded that of Newton, space agencies around the world, even now use only Newton's equations for their calculations as they have are quite accurate in the present scenario except for a minute difference while calculating the precession of planet Mercury. This leads us to a valid question, which is, if Sir Isaac Newton's theory of gravity is wrong, why are his equations so accurate. There are two possible explanations for this, which are: if gravity is not an attractive force emanating from the earth's center, then something must be pushing objects toward the earth from space or weight must be an inherent property of matter. Before we discuss more about these two possibilities, it is important that we first try to understand Einstein's theory of gravity as stated in the general theory of relativity.

Einstein's theory of gravity

In 1905, Einstein published his special theory of relativity in which he had created a new framework for the laws of physics. However, as STR (special theory of relativity) dealt only with inertial systems, it was incompatible with the gravitational force as described by Newton's law of gravity. This is what prompted him to develop the general theory of relativity.

Space-time

According to the general theory of relativity, gravity is not a true force but a curvature of space-time, and objects in orbit are not acted upon by any external force but are moving in their geodesics, which is the shortest distance between two points on a curved surface or trajectory. He explained this concept of space-time with the following analogy. If we place a heavy ball, representing the earth, on a trampoline as illustrated in fig. 30, it will cause a depression at the center. If a smaller ball, representing the moon is placed anywhere in the depression, it will fall back toward the ball at the center, that is, the earth.

Fig. 30

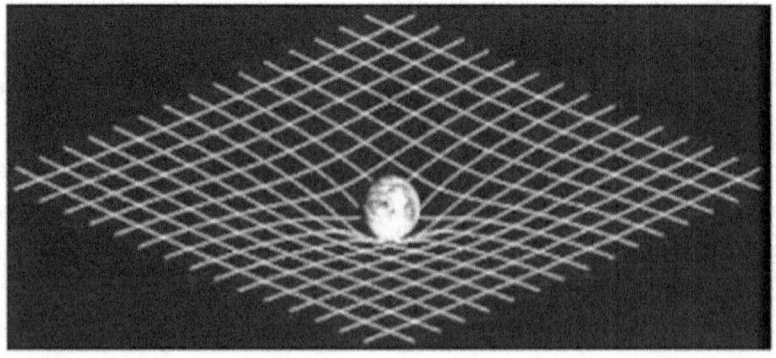

However, if it is moving around the bigger ball with the required angular velocity, then it will continue to remain in orbit. However, this illustration is two dimensional, and it is practically impossible to illustrate this concept in four dimensions namely three spatial and the fourth time. Even Stephan Hawking in his book, *A Brief History of Time,* has admitted that it is practically impossible to visualize anything as four dimensional.

In fig. 31, the central body is the earth, the smaller body represents a satellite in outer space, and the shaded portion around the earth is its atmosphere. Just as the earth's atmosphere curves around the earth, outer space also curves around it.

Fig. 31

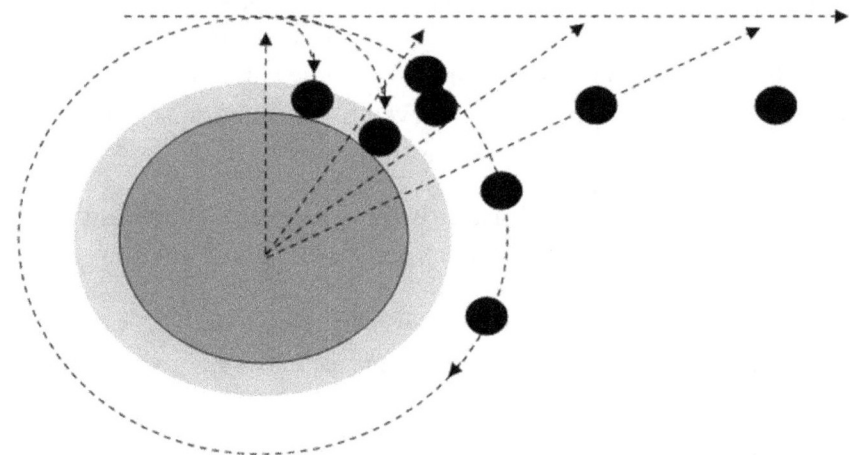

I always had difficulty in visualizing a curved vacuum but now it is so obvious. The earth's surface is curved and hence the vacuum of space also curves around it.

According to Sir Isaac Newton, a weightless object will travel in a straight line as illustrated in fig. 31 as long as no force acts upon it. That is why he presumed that an attractive gravitational force must exist between the earth and moon. He illustrated this with his famous cannon ball thought experiment in which the ball would continue to orbit the earth if it possessed the required orbital velocity. However, the reasoning that a cannon ball or satellite will continue to travel in a straight line in the absence of gravity cannot be true because for this we have to imagine that the earth's surface and subsequently space is flat. As the earth's surface is curved, in the process of traveling in a straight line, the cannon ball is also ascending with respect to the earth's surface, and as its mass is intact, it will lose momentum continuously due to inertia until at a particular altitude it may just stop moving or might fall back toward the earth. We must remember here that inertia of an object is entirely dependent on its mass and not on any direction.

According to Einstein's theory, no force is acting on the weightless satellite while it is orbiting the earth, but it is simply following its geodesic, the shortest distance between two points on a curved surface or trajectory, as space itself is curved. However, if this postulate is true, then satellites should be able to remain in orbit even at very slow speeds, which we know is practically impossible. If the speed of an orbiting spacecraft is reduced even slightly, it will begin to descend toward the earth. Hence, we have to conclude that a satellite travels in a curved path not because it is following its geodesic but due to some other reason.

There are only two possibilities by which a satellite can travel in a curved path in the absence of a gravitational force, which are as follows:

1) A satellite should possess weight, which is independent of gravity in which case the downward force arising from its weight will now do the work of gravity.
2) Its interaction with the medium pervading space somehow gives rise to a force that pushes it toward the earth.

Sir Isaac Newton himself posited the second possibility in his second edition of *Optics*.

Einstein's postulate that space is curved is absolutely true. However, the curvature is not due to space-time but primarily because earth's surface is curved and space also curves around it. As space is filled with a medium, this medium curves around the earth's atmosphere in exactly the same manner as the atmospheric air curves around the earth's surface.

However, as the theory of space-time curvature is still in vogue even 100 years after Einstein attributed the physical properties of space to ether and not space-time, it would be worthwhile to find out what space and time actually represent or mean. However, instead of digressing now, let us first study his equivalence principle and then revert to the topic of space and time.

Einstein's equivalence principle

Einstein had an inspiration in 1907, which he described as the happiest moment in his life, due to which he discovered that during free fall, objects experience weightlessness as they are falling along with the gravitational force. Based on this, he performed another thought experiment, which is as follows. If a spacecraft is accelerating upwards at 9.81 m/sec2 in a gravity-free environment in space, then a person standing inside it would experience a G-force of 1, the same as on the earth, and hence will be unable to distinguish whether she/he is inside the cabin of an accelerating spacecraft or standing on the earth's surface. From this, he concluded that acceleration due to gravitation felt on the earth is equivalent to acceleration felt inside an accelerating spacecraft and hence an independent gravitational force may not exist. According to his theory, a person standing still on the earth's surface is in an accelerated frame of reference, whereas an object in free fall or an orbiting body is in inertial motion. This is exactly opposite to what Newton's theory states.

However, as we have already discussed earlier, a man standing in a spacecraft accelerating upwards at 9.8 m/sec^2 will be unable to jump up as the speed of the spacecraft is constantly increasing. Only if the spacecraft is moving upwards at constant velocity, he may be able to jump up and rise above the floor. However, then he would not feel the G-force either. When he jumps up on the earth's surface, he can feel himself going up and then dropping down, as the earth's surface remains stationary relative to him.

Another important point that we should take note of is that the spacecraft is accelerating only because a force is acting on it, which clearly highlights the fact that acceleration can occur only under the influence of a force. Moreover, acceleration can be sustained for only a limited period of time as the amount of energy required will increase exponentially. In a spacecraft, the moment acceleration ceases, the G-force will also disappear, whereas in the case of the earth and all other celestial bodies, the G-force is felt perpetually even when the objects are in an inertial state.

The following is an excerpt from Wikipedia in which Einstein was criticized for exactly the same reasons by Lenard and Gustav Mei. "*They argued that physical forces can only be produced by material sources while the gravitational field that Einstein supposed to exist in an accelerating frame of reference has no concrete physical meaning. Einstein responded that based on Mach's principle, one can think of these gravitational fields as induced by the distance masses. In this respect, the criticism of Lenard and Mai has been vindicated, since according to the modern consensus, in agreement with Einstein's own mature views, Mach's principle as originally conceived by Einstein is not actually supported by general relativity.*"

It is now explicitly clear that although acceleration temporarily gives rise to an effect similar to gravity, the gravitational fields existing around celestial bodies cannot be considered equivalent to acceleration by any stretch of imagination. Similarly, the accelerating motion of bodies in free fall cannot be termed as inertial.

What space and time actually represent

Immanuel Kant, an eminent philosopher in his book, *Critique of Pure Reason*, opines that both space and time are not physical substances or entities in themselves.

Outer space

Before we begin our discussion on space, let us visualize a white board, which is completely blank. If we paint a beautiful picture on it, we will only observe the painting without thinking of the blank board. We can draw or write anything else on the board by wiping out the earlier picture and can repeat it any number of times. Similarly, we can hang three-dimensional objects on the board. However, every time we rub the board clean or remove the hung objects, we will again see the white emptiness. It makes no difference if we paint something on it or not the emptiness of the board will remain perpetually in the background. Now, if we disregard the physical nature of the board, then space is like the empty board, but three-dimensional in nature. Instead of the board we can also visualize an empty room. Whether physical objects exist in it or not, the emptiness will remain forever like the blank white board.

Earth's atmosphere behaves like a physical entity having elastic properties only because of the presence of air. Let us for a moment visualize the consequences, if all the air, water vapor, dust, and so on suddenly vanishes from the earth's atmosphere, will the space that contained the air and all the other particles also disappear? Definitely not, the emptiness will still exist. However, it will not possess any physical qualities, as it is now absolutely empty, a pure vacuum.

Likewise, if all the celestial objects existing in the universe including light and all the other electromagnetic waves, radiation, and other things were to suddenly vanish, the emptiness of space will still remain as it is, but will not possess any physical qualities.

Sometimes we come across articles in which it is said that space is expanding or being created and that before the big bang, there was no space. This cannot be true because absolute emptiness is not a physical entity. Logically speaking, emptiness is the natural state of the cosmos and cannot be created nor destroyed because it is not a physical entity by itself and if emptiness itself were not to exist, it is difficult to even visualize as to what would exist in its place.

Time

Quote by J.M.E. Mc Taggart, *"Our ground for rejecting time, it may be said, is that time cannot be explained, without assuming time."* According to him, time is just an illusion.

Birth and evolution of time

Our modern civilization would go into utter chaos if all of a sudden the concept of time and all the clocks, watches, and other devices, which are used to measure or indicate it, were to suddenly vanish because our entire lives are more or less governed by time, and it is of utmost importance in our daily lives. The funny thing about time is that there is a question about whether such a thing exists or not. Even the most brilliant minds in the scientific world and other eminent persons of society differ vehemently on the existence of time. Some are of the opinion that time really exists, whereas others believe that time does not exist; only watches and clocks do.

Let us try to find out the truth about time by finding out how, when, and why the present-day concept of time evolved.

For this, we have to start from the beginning of human civilization. As human beings evolved from bacteria over a period of millions of years, they began to socialize and the need for communication arose, which in turn led to the invention of languages. After further evolution, trade and commerce began to be practiced and a necessity arose for describing the size and heaviness of objects. The concept of length, breadth, height, volume, and weight along with mathematics came into existence and these parameters were used to describe the physical characteristics or size and shape of objects. Our ancient ancestors living in different places slowly started the practice of meeting one another for various purposes, personal as well as business. Now they had to communicate with each other as to when to meet. As there were no clocks in those days, they began to use the daily phenomena of sunrise and sunset as a form of time. They gave different names to different parts of the day and night starting from one sunrise to the next, like morning and early morning, afternoon, evening, night, and midnight and so on. These terms served the purpose of time in those days and were used to communicate as to which part of the day or night it was. Gradually, sundials, sand clocks, and similar devices were invented to measure time.

As we can see, the concept of time evolved based on the period between one sunrise to the next, which in turn depends on the rotation of our earth, that is, the period required for the earth to complete one full rotation around its axis.

As days, years, and centuries passed by, technology developed and the need for more accuracy also arose. This in turn gradually led to the invention of clocks. In the period between one sunrise and the next, earth makes one complete rotation about its axis. If we consider any point on the earth's surface, it would have rotated exactly by 360° and come back to its starting point. Each degree here represents 1/360th part of one full day. The first clocks had just a single hand, but as it was difficult to represent one full day or 360° in a single circle, some clever engineer or scientist divided one full day into 24 units, for reasons best known to him, and called each unit as 1 hour. Hence, one revolution of the earth is split up into 24 rotations of the hour hand in a clock, thereby increasing its accuracy. To further enhance the accuracy, each hour was then further divided into 60 minutes and each minute into 60 seconds. Clocks and watches from then onwards were manufactured with three hands.

As one full day is divided into 24 hours, 1 hour represents 1/24th part of one full day. If we visualize a full day only in terms of minutes, then one full day will consist of 60 into 24 hours, which is equal to 1,440 minutes. Therefore, 1 minute represents 1/1,440th part of one full day. Similarly, one full day in terms of seconds will consist of 86,400 seconds, and hence, 1 second represents 1/86,400th part of a full day. That is why a second, until recently, was defined as the fraction of 1/86,400 of a mean solar day. Now scientists have invented timepieces that can measure even a millionth of a second or more.

From the above discussions, it is explicitly clear that just as we use length, breadth, height, volume, and weight to describe the physical properties of an object, time was initially used to describe or indicate which part of a day or night it was. This in turn enabled people from different places to meet at any a particular moment. Time was also used to describe when certain events would occur or take place. This is the actual reason for the invention of time. However, we now use it for various other purposes; for example, stating the speed of light, acceleration, and so on

Because modern civilization is heavily dependent on this concept of time, it has assumed a unique status, which is next only to money and God. Just as people discuss the existence of God, so also scientists discuss and differ about the existence of time.

However, one thing is certain and that is, time is not a physical entity like radiation, electromagnetic waves, and so on, which although invisible can be considered as the energetic state of matter.

In this new perspective, let us visualize as to what would be the consequences if instead of considering that one full day consists of 24 hours, we assume that it contains only 12 or 48 hours.

If we assume that a full day consists of only 12 hours, then each hour will be equal to 2 hours of today's standard. The speed of light in this case would be equal to 600,000 km/sec.

If we assume that a day consists of 48 hours, then 1 hour will be equal to half an hour of today's standard. In this case, the speed of light would be 150,000 km/sec.

With our present standard of time, the speed of light is 300,000 km/sec.

It is interesting to note that the actual speed of light remains constant in all three cases. However, the figures or values change because of the use of different standards or interpretations of time and hence the duration of a second, minute, and hour in all these cases will be different.

Hence, as long as we use time for comparative purposes only, as is mostly being done today, there will be no problems. However, it may not be appropriate to use it in other contexts because this may lead to strange problems. For example, in the equation $e = mc^2$, the speed of light is directly used to calculate the energy equivalent of a given unit of mass. According to our earlier discussions, although the speed of light is a universal constant, the standard of time used by us is not a universal constant. Our time is based entirely on earth's rotation and we have already observed how the value of the speed of light per second changes depending on how many hours one single rotation of the earth is divided into. At present, it is just by chance that it has been divided into 24 parts and there is no solid scientific reason behind it. Just visualize what result we would get for $e = mc^2$ if instead of 24, a day had been divided into say 20 or 22 hours.

There would also be problems if we ever met with aliens in the future whose time may be based on the rotational period of their planet, which will definitely differ from earth's rotational speed. In such a scenario, it will be impossible to decide which value of the speed of light is to be used in Einstein's formula.

From our discussions, it is clear that everything in this universe will remain and function in the same manner as it is doing today irrespective of which standard of time we follow.

Some modern scientists have begun to realize that time is not a physical entity. In the words of Amrit Sorli and David Fiscaletti, from the scientific research center, Brista in Ptuj, Slovenia, *"time by itself has only a mathematical value, and no primary physical existence."* Kurt Gödel, a mathematician cum philosopher and a colleague of Einstein, "in 1949 produced a remarkable proof." *"In any Universe described by the theory of relativity, time cannot exist."* *"Our research confirms Gödel's vision; time is not a physical dimension of space through which one could travel into the past or future."*

Now it is time for all of us to introspect about the concept of time and answer the following questions on time; is time a physical entity or just a notion? Can space and time really intermingle or are they separate entities? Does time really exist and if yes, where?

While you figure out the answers for the first two questions, I will answer the third question.

Answer: Time is just a notion and exists only in the human mind.

It is now explicitly clear that, as opined by Immanuel Kant and numerous others, the emptiness of space by itself and time are not physical entities meaning that emptiness by itself and time are not made of particles like the photons, whereas the universe that occupies this empty space is entirely physical in nature. Even energy in the form of light and all the other electromagnetic waves are comprised of tiny particles called photons.

As we now have a broad idea about Einstein's theory of gravity, let us find out Newton's thoughts at the time he wrote the universal laws of gravitation. Instead of continuing with our preconceived notions, let us begin from the start with an open mind, just like a detective begins his investigation starting from the crime scene, and see if we are able to discover anything new, which might help us in solving the mystery of gravity.

Newton's reasoning while writing the universal laws of gravitation

Newton's original view was that because celestial bodies were considered weightless and space was an empty vacuum, a body in motion would always travel in a straight line when no other forces were acting on it. However, because the moon was revolving around the earth instead of going in a straight line, it occurred to him that an attractive force must exist between the moon and the earth. He also reasoned that if the moon's movement were stopped and if it was brought close to the earth, then it would also fall toward the earth at the same rate of acceleration as objects on the earth would. This led him to the conclusion that the moon and the objects on the earth were attracted by the same force, that is, gravity; if the objects on the earth were attracted by a different force or possessed weight on their own, then they would have fallen down with a greater rate of acceleration. Because of this reasoning, weight became a characteristic of gravity and the long-held notion that it was an inherent property of matter was abandoned.

Newton faced a major problem while formulating the theory of gravity and that was to do with the manner in which forces normally act. The total amount of pressure acting on an object is normally dependent on the surface area on which the force is acting. Newton cleverly overcame this problem by positing that the gravitational force acts in proportion to the mass contained in a body. Everyone still believe in this even though the mechanism by which an inanimate and unintelligent force can distinguish between a thin rod standing in a vertical position and a very thin sheet, of the same mass, resting on the ground and act accordingly is unknown. Logically speaking, no normal force can act in this manner as it involves judgment and discretion.

Newton's original concept of weight is entirely different from that which we follow today.

Excerpt from the book, *Excurse into the History of Weight* by Igal Galili and Michael Tseitlin.

"Following Newton's discovery regarding the nature of weight, as an interactive force of gravitation, weight ceased to be a characteristic of objects *while mass (the quantity of matter) remained such. The often forgotten feature of Newtonian weight was however, that it always came as a pair of forces of interaction. Newton wrote, 'The weight of the planets toward the sun must be as their quantities of matter.'"*

This meant that the weight of the earth toward the sun was equal to the weight of the sun toward the earth, and the weight of the earth toward the moon was equal to the weight of the moon toward the earth, and the weight of the earth toward the sun was different from the weight of the earth toward the moon. The Newtonian weight was not the characteristic of a body but a pair of bodies. Such weight could not survive in the everyday life, where the only practical meaning was the weight of things toward the earth.

As Newton's theory of gravity begins from the notion that all celestial bodies are weightless, let us try to find out if this is true.

Celestial bodies and weightlessness

This is a very important topic because Newton's theory of gravity is based entirely on the assumption that matter by itself is not heavy.

As we have already discussed about this subject in chapter one, let us briefly recall the reason why objects entering space appear to become weightless.

In outer space, the weight of matter contained in a celestial body will always act toward its own center of mass. As there are no upward or downward directions in space, for the body as a whole, all directions are the same and hence when it is far away from the gravitational field of other bodies, where space is isotropic, it will float giving rise to the illusion that it is weightless. This observation is very important because even if weight were to be an inherent property of matter the result would still be the same.

This also reveals another interesting aspect of space, which is that every conceivable point in space behaves as if it is the center of the universe because from that point every direction is upwards. From the point of view of a celestial body located anywhere in space (e.g., earth), all directions away from its center of mass are upwards and toward it are downwards.

However, this will be true only within its gravitational field only. Outside this boundary downwards is toward the sun.

Newton's views of gravity in his earlier and later years

Although Newton, in his younger days, stated that the gravitational force between two bodies was proportional to the product of their masses, he was unable to explain the exact manner in which mass gave rise to the gravitational force. Hence, he proposed no hypotheses to explain the same as is evident from the following quote, *"You sometimes speak of gravity as essential and inherent to matter. Pray do not ascribe that notion to me, for the cause of gravity is what I do not pretend to know, and therefore would take more time to consider it."*

Even after Newton's gravitational theory was accepted, he was constantly criticized by a few of his contemporaries for not explaining how gravity was able to act across large distances instantly through the vacuum of space. This fact bothered him throughout his life due to which as he grew older, he proposed a few mechanical ether theories of gravitation of which the following theory published in his second edition of *Optics* (1717) is the last and is the most probable cause of gravity. *"Is not this medium (ether) much denser within the dense bodies of the sun, stars, planets and comets, than in the empty celestial spaces between them? And from passing from them to great distances, doth it not grow rarer and rarer perpetually and thereby cause the gravity of those great bodies toward one another and of their parts toward their bodies; every-body endeavoring to go from the rarer parts of the medium toward the denser? And though this decrease of density may at great distances be exceedingly low, yet is the elastic force of this medium be exceedingly great, it may suffice to impel bodies from the rarer parts of the medium toward the denser, with all the power which we call gravity."*

From these words, it is clearly evident that an old but more mature and wise Newton had changed his earlier views on the cause of gravity. According to his letter written to Robert Boyle, which we have already touched upon in an earlier chapter, celestial bodies do not attract each other but are pushed toward each other by the all-pervasive mechanical ether, which is endowed with very powerful short-range repulsive forces.

This new postulate not only provided a logical explanation for the long range of gravity but also did away with the earlier spooky action at a distance theory.

This postulate is exactly similar to the contemporary gravitational field theory in which, however, the ether has been wrongly discarded. It is an irony that his contemporaries rejected this particular theory, which is more logical than the earlier one, as he was unable to explain the density gradient of the ether. Because this theory is not taught in educational institutions, most of us are unaware of its existence, and hence even after 300 years, we still believe that gravity is an independent attractive force when Newton himself, who can be referred to as the father of gravity, had abandoned that idea long ago. This is why the mystery of gravity remains unresolved even until today.

The following is the opinion of Eric Baird, a scientist, who has written an article about the various mechanical ether theories of gravitation proposed by Newton. *"Newton's ether model arguably represents one of the most serious missed opportunities in the history of gravitational physics."*

Leonhard Euler's views on gravity

The following is Leonhard Euler's, a mathematical genius and a philosopher, hypothesis regarding gravity, which conforms to Newton's later views.

"Gravity is the external cause, which forces terrestrial bodies downwards; and therefore it cannot be a property assigned to certain bodies themselves. What is more likely to be true is that the force of gravity arises from the action of some more subtle matter that escapes the notice of our senses. The subtle celestial air (ether) is in a forced state, and is compressed far beyond its natural density, for which reason it exerts everywhere an unusually strong spring force and compresses all bodies."

Let us now study the reasons that favor and do not favor the existence of an independent gravitational force.

Points that support the existence of an independent gravitational force

1) Celestial bodies seem to obey the universal laws of gravitation, which states that gravity between two bodies is directly proportional to the product of their masses and inversely proportional to the square of the distance between them.
2) Most people believe that most celestial bodies are in the shape of a sphere because of gravity.
3) The third reason has to do with Newton's second law of motion, which states that an object will accelerate only when acted upon by an external force. As freely falling bodies in outer space and on the earth's surface accelerate, gravity is the external force.

Points that do not support the existence of an independent gravitational force

1) Although gravity is supposed to be millions or even billions of times weaker than the other fundamental forces, its vast range defies all logic. Similar to the strong nuclear force, it totally violates the observed natural laws of nature where the range of a force increases with increase in its strength and vice versa.
2) It is always attractive and has no opposite force. Most other forces come in pairs where opposite forces attract each other and like forces repel.
3) Although the net pressure exerted on an object by forces other than gravity depends on the surface area of the object the gravitational force depends entirely on the mass or the matter contained in objects. This means that an object having a given mass can be in any shape or form, for example, in the form of a sheet, wire, ingot, or rod, the gravitational force will somehow detect this and hence ensure that the object's weight always remain constant. This is in a way suggesting that gravity is an intelligent force.
4) It cannot be shielded against because it is a force that can penetrate through solid, liquid, and gaseous matter including huge celestial bodies like planets, stars, and so on without the slightest diminution

in its strength and at the same time endowing matter with weight. It is hard to visualize how such a weak force is not only able to traverse vast distances but also penetrates and passes through thousands of kilometers of dense matter. Moreover, if it has to induce weight, it should encounter at least some resistance while passing through matter in which case its strength should reduce proportionately.

5) A force such as gravity, which causes objects to accelerate, requires a constant source of a mind-boggling amount of energy to give rise and sustain it, and if this energy is originating from our earth, then the earth's mass should constantly be decreasing. However, we know that the earth's mass is more or less constant.

6) If the weight of objects was indeed due to gravity alone, then it should have been detected a long time ago. A force requires a force carrier, and according to the Standard Model in physics, gravitons are the force carriers in the case of gravity. However, even though particles like the neutrinos, which are neutrally charged and are thousands of times smaller than electrons, have been detected, gravitons have not yet been detected.

7) Solid and liquid matter at the microscopic (atomic and molecular) level is held or bonded together by the inter- and intramolecular forces, and hence, as far as objects themselves are concerned, irrespective of their mass, a gravitational force is not required to hold bodies together. Voyagers 1 and 2, space probes launched by NASA in the year 1977, are so far away that although in a few more years, they will be out of the sun's gravitational field they have not disintegrated. Gravity seems needed and exists only in outer space to enable celestial bodies to continue to orbit.

8) As the strength of gravity is negligible at the atomic, molecular, and subatomic levels, what is preventing the air in earth's atmosphere, which exists as individual atoms and molecules from escaping into space?

9) During a solar eclipse, the moon partially blocks the sun from the earth's view for a few minutes by coming in between them. In this brief period, if we go according to the prevalent theory, then the gravitational interaction between the earth and the sun should also be blocked at least partially. However, we are all aware that the earth is unaffected

during this event, which is borne out by the fact that it continues to move in its orbit without the slightest change. As gravity is supposed to be a force that acts in between the center of mass of two bodies, the moon's presence should have decreased the gravitational attraction between the earth and the sun. However, instead of a decrease, there seems to be an increase in gravity due to which the ocean tides on the earth get bigger than at other times. The moon also continues to orbit as usual, in this brief moment, even though the sun and earth's gravity is pulling it in opposite directions.

10) The tides that occur on earth are attributed to the attractive force exerted, largely by the moon and to a lesser extent the sun, on to the earth's oceans. If this is true, then the moon's gravity should also be attracting the other parts of the earth including the objects that are not rigidly attached to the ground because of which the weight of these objects should vary depending on the moon's position. For example, the weight of unattached objects that happen to be directly below the moon should be slightly lesser than the weight of objects present on the other side of the earth. Similarly, as the distance between the earth and the moon varies in the course of the moon's orbit around the earth, the weight of unattached objects should also change accordingly. As we now have digital weighing machines that are highly accurate, they can measure with an accuracy of one-thousandth of a milligram or even more, it should be possible to record even the slightest change in weight. However, in practice, it has been found that neither the position of the moon and the sun nor the distance between them and earth in the slightest bit affects the weight of unattached objects on the earth.

11) Objects ascending above 100-km altitude from the earth's surface are subjected to microgravity, which literally means one-thousandth of the gravity that exists on the earth's surface. However, according to the calculations based on the inverse square law, the strength of gravity should decrease gradually and not suddenly. In between 100- and 400-km altitude, its strength should still be about 97%–90% of its value on earth. The huge and sudden drop in the strength of gravity above 100-km altitude does not support the theory of an independent gravitational force.

12) All celestial bodies, for example, planets in our solar system, have a gravitational field, the extent of which is supposed to depend on their as well as the sun's mass and the distance in between them. This sphere of influence in which stable orbits are possible is known as the Hill Sphere. However, according to the information gathered from Wikipedia, this value is only an approximation as it has been practically established that stable orbits are possible only within half or one-third of the derived value. Earth's Hill Sphere is around 1.2 million km, whereas Pluto's Hill Sphere, whose mass is about one-tenth that of earth, is supposed to extend up to 9 million kilometers. The recent flyby of the New Horizons mission across Pluto establishes the fact that its Hill Sphere is far lesser than the calculated value.

13) Michael Faraday was the first person who illustrated how the electromagnetic field lines originate and flow in a loop with the help of iron filings. However, nobody has yet tried to map the gravitational field, even theoretically and illustrate how its field lines flow because its origin and exact nature of action are still a huge mystery.

14) All of us have learnt that the strength of gravity on the moon's surface is six times lesser than the strength of earth's gravity. This means that an object weighing 360 Newton on earth, will weigh only 60 Newton on the moon. NASA has successfully sent six manned missions to the moon and one would have expected their astronauts to conduct a weighing experiment to confirm the above. Newton's laws of gravitation have so far been confirmed only by celestial observations by studying the motion of celestial bodies and so on. Here were golden once in a lifetime chances to conduct a simple weighing experiment on the moon's surface and further confirm the validity of Newton's laws of gravitation. This would have become the first terrestrial confirmation of Newton's laws of gravity. However, although this experiment was and is even today of vital importance, no information is available whether such an experiment was ever conducted, and if it was what were the results. It is truly strange if they did not conduct such an experiment, which although simple is of vital importance.

After the Americans, the Russians and the Chinese have also managed to soft land vehicles on to the moon. However, even they have not released any information as far as the actual weight of objects on the moon's surface is concerned.

There can only be one reason behind this mysterious lack of information on this subject and that is the results obtained did not conform to what was expected. Newton's laws of gravitation seem to work perfectly only in space.

If the strength of gravity on the moon's surface is truly one-sixth the strength of gravity on the earth, then Newton's laws of gravitation are perfectly validated. In that case, why do modern scientists believe that Einstein's gravitational theory has superseded that of Newton?

All the above points strongly imply that gravity is not an independent attractive force emanating from the earth or any other celestial body.

Experiments, which prove that an independent gravitational force does not exist

Thought experiment number 1

According to the existing theory, weight is a property endowed by gravity, and hence, in the absence of an attractive gravitational force, it is hypothesized that all unattached objects would float away into outer space. This concept is illustrated in fig. 32, where an object is resting on the earth's surface. Mass, on the other hand, remains constant throughout the universe. Mass is also referred to as inertia and inertia itself is defined as a body's resistance to motion.

Fig. 32

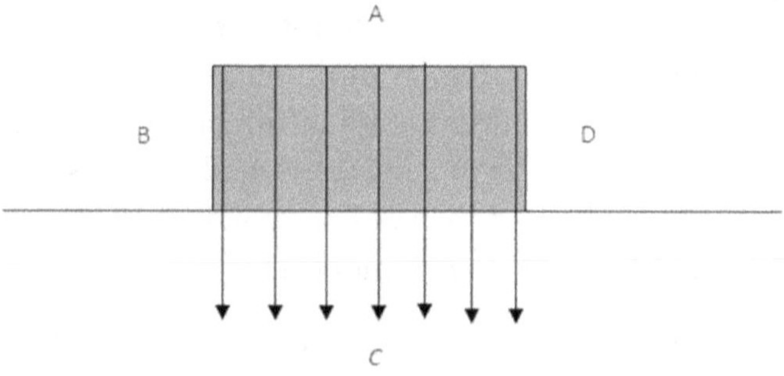

Although this reasoning seems logical, there is a valid doubt, which needs clarification. Just as gravity always acts in the vertical plane, is inertia felt only in the horizontal direction? For example, in fig. 32, weight is acting across sides AC and inertia is felt across sides BD. According to the current theory, if the object is now turned so that surface B comes on top weight will be felt across sides BD. Does this mean that the inertia felt earlier across BD has vanished and is now felt across sides AC? Logically speaking, inertia must be felt from all directions because it is an inherent property of matter. Hence, even in the absence of a gravitational force, an external force has to be applied in the upward direction to overcome inertia and set it in motion.

Thought experiment number 2

If a small rubber ball tied to a thin elastic band is thrown in the horizontal direction, we will notice that as the ball moves away, the elastic band stretches, and after traveling a certain distance, it is pulled back by the stretched elastic band. The ball is pulled back because the kinetic energy that entered the system is stored as potential energy in the elastic band as it stretches out, and when the band contracts, it is again converted back to kinetic energy.

Now let us observe an object, which is thrown vertically upwards. We will notice that first it will decelerate until it reaches a certain height after which the entire process is reversed. It will accelerate at 9.81 m/sec^2 as it falls down until it crashes onto the ground. Both, deceleration and acceleration, is attributed to the pull that the gravitational force is exerting on the object. Once again, an interesting question needs an answer. Is the freely falling object accelerated by the pull of gravity or by the potential energy stored in it? The ever-attractive gravitational force that is supposed to be perpetually acting on the object is itself capable of forcing the object to fall and accelerate. Similarly, as we have observed in the example of the rubber ball and elastic band, the potential energy stored in the object by virtue of its distance from the earth's surface is also capable of bringing the body down. According to contemporary theory, objects fall down because of the potential energy stored in them and this is referred to as the gravitational potential energy. Here, gravity is compared to the elastic band, which is highly questionable because unlike an elastic band it is not a physical entity, which can store energy from an external source but is itself a force. We cannot also assume that the gravitational force alone is forcing the object to fall down because then the potential energy stored in the object will go unaccounted for thus violating the law of conservation of energy.

Newton applied the following logic while writing the gravitational laws. "The same force that acts on the moon is also acting on bodies on the earth because if gravity was different from this force, then bodies making for the earth by both forces acting together would descend twice as fast." If we analyze according to this logic, then the rate of acceleration of an object in free fall should be twice the normal value because two forces, the gravitational force and potential energy, are now acting upon it. However, as it is falling with the normal rate of acceleration, it is explicitly clear that only one force is acting on it and that is the force arising out of the stored potential energy. This reasoning forces us to conclude that an independent gravitational force does not exist at least on the earth's surface.

Thought experiment number 3

Let us visualize a huge chimney in which air is continuously sucked from the bottom, giving rise to an air draft flowing from the top to the bottom. An object, which is resting on a weighing scale inside the chimney, is acted upon by this continuous draft of air, which represents the gravitational force, due to which the scale registers a particular reading. Now let us find out what will happen if we move the object along with the scale upwards with constant speed. Will the object's weight increase or will it remain unchanged? It will definitely increase even though it is not accelerating because it is moving against the force of the air draft. Likewise, even if it descends with constant velocity, its weight will show a decline.

Now let us visualize an object that is placed on a weighing scale resting on the floor of an elevator that is ascending rapidly at a steady speed. Will its weight also increase? No, the object's weight will not increase. It will increase only if the lift begins to accelerate. From this, it is clear that a gravitational force does not exist because if it was really present, then just as in the case of the object, which moves against the air draft, the weight of the object inside the elevator should also have increased.

We can draw the following conclusions from this thought experiment:

1) When an object experiences weight because of an external force acting on it, then even the slightest and also uniform movement against that force will lead to an increase in its weight. At higher speeds, the increase in weight will also be greater and vice versa.

2) When an object experiences an increase in weight only when it is accelerating in the upward direction and vice versa, then it can be stated with certitude that the weight experienced or possessed by the object, when at rest, is solely due to the heaviness of the matter contained in it.

Thought experiment number 4

Einstein first postulated the relationship between acceleration during free fall and weightlessness in the year 1907.

He realized that when a person falls freely from the top of a tall building, in a gravitational field, he will be unable to feel his own weight because he will now be falling along with the gravitational field. In other words, he will experience weightlessness during the entire process of free fall. By borrowing the same analogy and with a little improvisation, we can decisively find out if an attractive independent gravitational force exists.

Let us visualize that we are standing on a weighing machine, which is resting on the floor of a lift of a very tall building and the lift itself is on the topmost floor. The weighing machine will indicate our weight as X kg. Let us also visualize that a small bucket containing water is resting on another weighing machine, which is also resting on the lift's floor and its weight is indicated as Y. If the cable supporting the lift were to snap suddenly, then the lift along with everything inside will begin to fall toward the earth with an acceleration of 9.81 m/sec^2. The weight indicated on the weighing machine will now be zero, and although we may feel a sense of weightlessness, we will not begin to float inside the lift. Similarly, though the weight of the bucket with water also registers zero, the water will continue to remain inside the bucket and will not rise up and collect itself into the shape of a sphere as it happens inside spacecrafts that are in space.

Now let us ask ourselves an interesting question. If we try to lift the bucket of water, whose weight according to the scale is now zero, to a certain height, should we apply force or will the bucket rise up effortlessly? If we analyze according to Newton's earlier theory of gravity, then we should effortlessly be able to lift up the bucket of water because according to his theory, the weight of an object, which is in free fall, will be zero. That is why it is universally believed that if gravity is absent on the earth's surface, all unattached objects would float away into space.

However, according to the results of the practical experiments conducted, we will definitely have to apply force to raise the bucket to the desired height and hence will feel the full weight of the bucket and water as long as it is rising upwards. However, as soon as we stop applying force, we will not feel its weight and it will appear to float if we release it as it is still falling with the same rate of acceleration as the lift. If we drop any object that is in our pocket, for example, a pen, we will notice that it will also remain suspended in the air. Weighing machines do not register our weight during free fall simply because for weight to be registered both the object as well as the support on which the weighing machine is placed on should be in a state of rest.

For arguments sake, let us assume that we are feeling the weight of the bucket filled with water in the falling lift because everything inside the lift is in a state of inertia with respect to each other. This argument will apply only when we are moving in a horizontal direction, like in a bus, train, or an aircraft and not in a freely falling lift accelerating at 9.81 m/sec^2 because the gravitational force is supposed to be acting on all objects in the vertical plane.

This experiment can be practically conducted with real people inside a module that is connected with parachutes and dropped from a suitable height with the help of an aircraft or helicopter. I sincerely hope that some dare devil stuntmen conduct this experiment in the interest of physics as the results are of fundamental importance and will once and for all either confirm the existence or shatter the myth of gravity. It can also be conducted in the various drop towers existing throughout the world with the use of cameras and suitable experimental material. However, I conducted this experiment on numerous occasions in the following manner. It is inexpensive and can be conducted by almost anyone.

Details of the practical experiment conducted

Items required

1) An iron block weighing about 700 g.
2) An electronic weighing scale.
3) A permanent magnet measuring 10*20*40mm
4) A hollow cube made of 5-mm acrylic sheet measuring 150 × 250 × 250 mm. Top should be detachable. Note: The size of the box depends on the size of the weighing scale.
5) A pyramid made with 5-mm acrylic sheet with a base dimension of 150*250 mm and height 180 mm.
6) A camera with a video recording facility.
7) Nylon rope of 6–7 mm diameter and length 15 m.
8) Necessary hooks and miscellaneous items.

Experiment details

The pyramid, which serves the purpose of reducing air friction during free fall, was attached to the acrylic box as illustrated in fig. 33.

Fig. 33

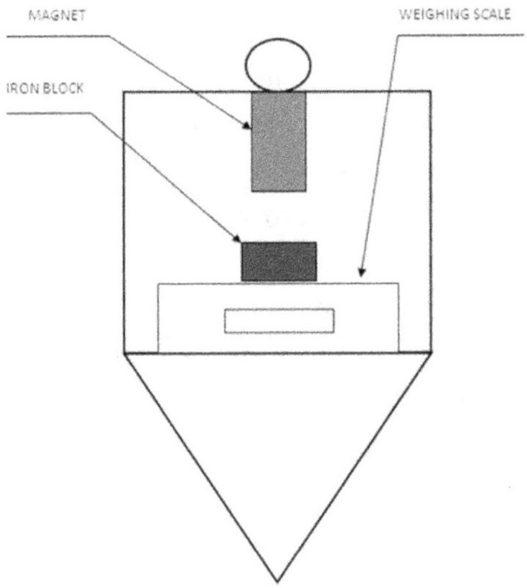

The iron block was placed on the base of the inverted pyramid and the whole contraption was dropped from a height of about 4 meters a number of times.

Results of the experiment

On every occasion during free fall the iron block was always resting on the base of the pyramid and did not rise up to float inside it. I also conducted this experiment in another manner. I attached a permanent magnet to the top plate and rested the block on an electronic weighing scale as illustrated in figure 33. When the top cover was swung into position the distance between the top portion of the block and the bottom portion of the magnet was about 25mm. The scale now indicated the weight of the box as 600 g instead of 700 g because of the magnets pull. Even when this contraption was dropped, I noticed that the block was immediately pulled up by the magnet and got attached to it. The result of this experiment first led me to believe that the block was really becoming weightless during free fall and hence weight was indeed a property endowed by gravity. However, on further contemplation I realized that this was not true. What was really happening is that the block was not falling simultaneously along with the contraption but a fraction of a second later due to the upward pull exerted by the magnet. If we consider this fraction to be just even $1/20^{th}$ of a second the magnet would have moved down by almost .49 meter in that time due to which the block gets attached to it.

Eventually, I realized that the perfect way of conducting this experiment was in a drop tower where the permanent magnet should be substituted with an electromagnetic one. It should be able to lift at least half the weight of the block when at rest. This electromagnet should be switched on only when free fall has already begun. If weight is truly a property endowed by gravity then the block will rise up and attach itself to the magnet as its weight would now be zero. However, in case this does not happen we will know for sure that weight is truly an inherent property of matter as was believed prior to Newtonian era.

Experiment 2

In this experiment, the weighing scale, magnets, and iron block were replaced with a flat iron bar pivoted at the exact center with a metallic block W suspended at one end as illustrated in fig. 34. If weight is truly a property endowed by gravity, then during free fall the weight of

Fig. 34

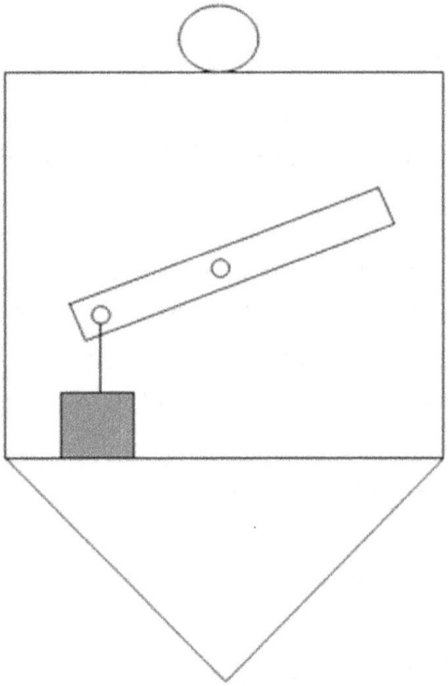

the block along with the bar should become zero due to which the inclined bar along with the block should assume a horizontal position. However, if weight is not a property endowed by gravity but a natural trait of matter, then the bar along with the weights will remain in the same position even during free fall.

Results of the experiment

After dropping the contraption a number of times, it was found on all occasions that the bar along with the weight remained in its original inclined positions.

Conclusion

The result of these simple experiment along with the other thought experiments irrefutably corroborates Einstein's and Newton's later postulate that gravity is not an independent attractive force that emanates from within celestial bodies and that weight is an inherent property of matter. The gravitational force has not been detected simply because such a force does not exist.

Abbreviations

Before proceeding to the new theory of gravity (NTG), specific names and abbreviations to the different entities that will be discussed are provided here.

1) Matter as usual refers to the protons, neutrons, and electrons and represents the energetic part of matter.
2) Second matter denoted as SM refers to the elastic low-density substance surrounding the nucleus of an atom and represents the physical part of matter.
3) Matter plus second matter as exists in the atoms is named as gross matter (GM).
4) The dilute form of the SM existing in space is named as gravither. This name is derived by combining the words graviton and ether. According to the Standard Model in physics, graviton is the mediating particle responsible for the gravitational force. Ether is the medium that was thought to pervade the whole of space through which the gravitational force was transmitted.

New theory of gravity

"Truth is ever to be found in simplicity and not in the complexity or multitude of things."

Sir Isaac Newton

I am referring it to as a theory and not hypothesis because it is based entirely on Newton's mechanical ether theory of gravity published in the second edition of *Optics* in the year 1717.

As both Newton and Einstein's theories do not explain the exact manner in which mass affects and gives rise to the gravitational force or field, the new theory begins by taking into consideration the atoms because celestial bodies themselves are nothing but a huge conglomeration of atoms. Hence, logically speaking, there must be a definite connection between the atoms present in a celestial body and the gravitational field. We now know from previous chapters that the empty space in atoms is not truly empty but contains a condensate of the extremely low-density elastic medium pervading the continuum of space, which we have named as gravither.

The following are the main features of NTG

1) An independent attractive gravitational force does not exist anywhere in the universe and hence weight, as was believed in the pre-Newtonian era, is a natural trait of matter implying that mass and weight are the same thing. That is why the inertia of an object is directly proportionate to its heaviness. It increases as the heaviness (weight) of an object increases and vice versa.

2) A gravitational field always arises and exists only around bodies that are in outer space (celestial bodies) because the mass of these bodies displaces the gravither present in space just as an object on earth's surface displaces the liquid into which it is immersed. Hence, an object is surrounded by its own gravitational field only when it is in space and not when it is in earth's atmosphere, that is, below 100 km from its surface. As posited by George Stokes and many other scientists, the gravither as a whole is stationary. However, the gravither that makes up the gravitational field of celestial bodies will behave as an integral part of that body and move along with it in the same manner as the atmospheric air and all other atmospheric layers.

3) As the gravitational field arises due to the displacement of the gravither present in space, its range is not infinity. There is a definite distance

starting from the surface of celestial bodies, which depends on their mass, where the gravitational field will cease to exist.

4) The manner in which gravity works on the surface and in the inner atmosphere of celestial bodies is completely different from how it works in outer space. When objects are within the inner atmospheres of a celestial body, their weight always acts toward the center of mass of that body. However, as soon as they enter space, irrespective of their mass, they behave as independent celestial bodies and their entire weight will now begin to act toward their own center of mass because of which they appear to lose weight. Their gross weight, which decreases as the altitude from the primary body's surface increases, will however act toward the center of mass of the primary body.

5) As stated earlier, weight is an inherent property of matter and is far higher than what is actually measured on the earth. The combined weight of the nuclei of atoms contained in a celestial body always acts toward the center of mass of that body. However, the second matter present in atoms always tend to rise upwards to join the gravither present in space, like an air bubble rises up in water to join the atmospheric air, due to which an upward force arises to counter the downward force exerted by the atomic nuclei. When the mass of a celestial body increases, its average density and weight also increase because a part of the second matter present in the atoms near the core of the body is squeezed out and escapes into space as gravither because of which the nuclei of atoms regain a portion of their original weight. From this, we can say that the nuclei and the downward force associated with them represent gravity, and the second matter and the upward force associated with them represent antigravity. Fig. 35 illustrates this concept.

Fig. 35

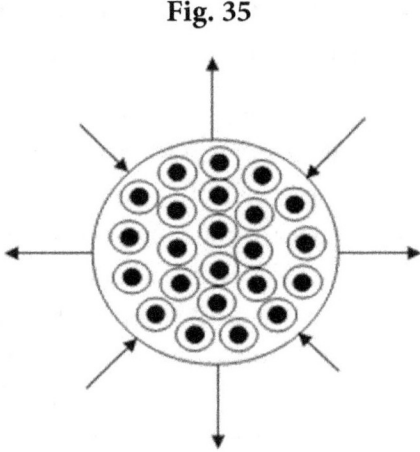

6) The GM contained in celestial bodies also behaves like matter and hence, just as in an atom, the GM and the gravither present in space will mutually repel each other due to which a compressive force begins to act all around the body. However, the intensity of repulsion will be far less than that within an atom. This force also contributes to the weight of matter and objects on the body's surface. Hence, two forces give rise to an object's weight. One is the downward force arising from the heaviness of matter and the other is the downward force exerted by the gravither membrane. Fig. 36 illustrates this concept.

Atom to earth size

Formation of the earth and its atmosphere

The human brain is the most advanced entity that has evolved and exists in the universe and one of its functions is imagination. In the words of Einstein, *"Imagination is more important than knowledge for knowledge is limited but imagination encompasses the whole world and leads to progress in science."*

Let us also make use of our imagination, by playing GOD, and visualize a single atom of any element, which is in outer space, and observe the various changes that occur in the body that will gradually come into existence and grow bigger and bigger as we continuously add more atoms. Let us assume that an infinite number of atoms are at our disposal.

As we regularly add atoms that are in a free state to the single atom, a tiny spherical blob of matter will first become visible. As we keep on adding atoms to this blob, the collective weight of the atoms will begin to act toward the center of mass of our imaginary body and compress the atoms near its core. Because of this, the average density of the body will steadily increase, whereas its volume will not increase in proportion to the combined volume of the atoms in a free state that constitute it. At the same time, the increasing mass of the body will steadily displace the elastic and motionless gravither present in space. This process will go on as we continue to add more atoms until the imaginary body attains a particular size or mass, when the compressive force exerted by the displaced gravither present in space along with the massive increase in weight acting toward the center of mass of the body becomes so strong that a portion of the SM present in the atoms that are near the core is squeezed out of the body. Due to this displacement of the SM, the body's average density and weight will further increase substantially.

Formation of the gravither membrane

The displaced SM immediately transforms into gravither and joins the gravither, which is already present in space to form thin elastic-like membrane surrounding the entire surface of the body. However, due to the mutual repulsion between the gravither membrane and the newly formed body, a gap of empty space, which represents its atmosphere, will arise in between the two. This is exactly similar to what happens within an atom.

As illustrated in fig. 36, the body is now surrounded by the repulsive gravither membrane and a weak gravitational field extending away from it. This gravitational field gradually comes into existence as the increasing mass of the body pushes the gravither membrane and the gravither present above it outwards causing its density to increase in a gradient manner.

Fig. 36

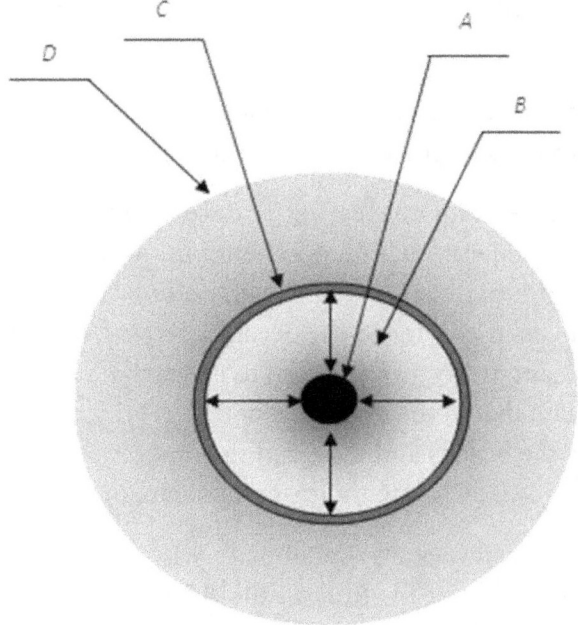

A - Earth

B - Earth's atmosphere containing air

C - Gravither membrane

D - Outer space consisting of second matter whose density decreases as the distance from earth increases

Let us now assume that the mass of this imaginary body keeps increasing until it attains the mass of our earth. The intensity of the two forces mentioned previously and which are acting toward the center of mass of the body would have also increased substantially. Because of the increase in the intensity of these forces, all the GM present near the earth's core is compressed to such an extent that a lot more SM escapes from the body as gravither into space. The average density and weight of GM contained in it would have further increased due to reasons already mentioned.

If the earth did not have an atmosphere, then the highly elastic gravither membrane would have formed much closer to the earth's surface because, then, only the repulsive forces between GM and gravither would have come into consideration. However, as we do have an atmosphere, this membrane will form high above it, as air is also a form of matter both will mutually repel each other.

The vicissitudes of the earth's surface are superimposed on to the gravither membrane and gravither present in space.

As the distance from the earth's surface increases, the unevenness gradually reduces and eventually smoothens out. As all the matter on earth has now regained more of its actual weight, all objects, including human beings and animals, the water in the oceans and seas, and the gases in the atmosphere, remain firmly grounded on to the earth's surface. It is our own weight (i.e., weight due to the heaviness of matter) that is keeping all living beings, water in the oceans and seas, and the other movable objects, which are not permanently attached to the ground, to remain on the earth's surface.

Most of us have always wondered as to why 80% of the gases in our atmosphere remain confined within the first 10 kilometers above the earth's surface and not spread out evenly in the balance 90 km's of empty space before outer space begins. The reason for this is of course gravity.

The only difference now is that there is no force emanating from within the earth that is preventing the atmospheric gases from escaping into space, but the repulsive gravither membrane existing above the atmosphere is pressing the gases down onto the earth's surface. The strong repulsion between the gravither membrane high up in the sky and the atmospheric gases (which is matter) gives rise to atmospheric pressure.

Objects in earth's atmosphere fall down and accelerate mainly due to their own weight and the potential energy stored in them. The repulsive gravither membrane also gives that extra push, which makes the falling objects to accelerate slightly faster.

If we now consider the weight of an object on the earth's surface, it will be equal to the sum of two forces. The first force is of course the force arising from the heaviness of matter, which will constitute more than 90% of the objects weight, and the second force that acts on the object from space is the force arising from the mutual repulsion between the gravither membrane and the object. As the mass of a heavenly body increases, the intensity of the second force will also increase. Hence, the ratio of its contribution to an object's weight on a heavenly body's surface will steadily increase as the mass of a heavenly body increases and vice versa.

The mutual repulsion between the GM contained in earth and the gravither present in space has actually created a bubble in the ocean of space inside which the earth is suspended. We can safely presume, by comparing data from other planets, that the distance that the second matter is pushed up is proportional to the mass of a celestial body. As the mass of a body increases, this distance should also increase and vice versa. In the case of the earth, it is 100 km, whereas on the moon, it is only about 20 km or so approximately. We can infer this easily by observing the altitude above which weightlessness begins and stable orbits are possible. On the moon, man-made satellites are even now orbiting at an altitude of about 25–30 km only.

At normal speeds, objects are able to pass through it without any visible signs of friction or drag. However, when objects travel at very high speeds, then friction develops due to which the objects heat up. It is because of the presence of this membrane that spacecrafts re-entering earth's atmosphere bounce back into space at exactly 120-km altitude if their AOA is more than 40° with respect to the vertical plane and heat up tremendously if they manage to penetrate it. The speed of the spacecraft during this period is about 27,000 km/hr.

Functions of the elastic gravither membrane

To find out the function of this membrane, let us for a moment observe the world around us. We will notice that right from bacteria to plants, trees, fruits, vegetables, animals to humans all are covered with skin, which protects the delicate internal matter from coming into direct contact with the harmful external environment. Even on inanimate objects like metals, rocks, and so on, a thin layer of oxidation acts as a barrier. The gravither membrane acts like the skin of our earth and protects everything on its surface and inner atmosphere from the harmful effects of outer space. Let us now visualize a submarine that is moving in the deep oceans of the earth.

As long as a sailor is within the pressurized submarine, he is able to breathe and move about. However, as soon as he goes out into the ocean, he will begin to experience neutral buoyancy or weightlessness. Moreover, he will not be able to survive for long, as water is not a suitable environment for humans. The submarine's outer shell prevents the air from escaping outside as well as the water from entering inside it. In a similar manner, when a spacecraft enters the environment of space, its outer shell not only prevents the air from escaping out but also acts as a barrier between the interior of the spacecraft and the harsh environment of outer space.

Due to some material defect, let us assume that an accident tears a hole in the outer shell of the submarine. We will notice that in an instance the air will escape outside allowing the water to enter inside the submarine. In the case of the spacecraft, the air will escape into space exposing its interior to the harsh environment of outer space.

If the earth is viewed as a whole, we will notice that it is exactly similar to a spacecraft traveling in space except for the fact that when compared to the spacecraft, it is very huge and has no rocket engines. Although our earth and all other celestial bodies are indeed natural spacecrafts, the same rules that apply to a spacecraft will also apply to them. If there is no barrier between the earth's surface and space, then just as in the case of the spacecraft, which developed a hole, the entire air present in the earth's atmosphere will disappear into space. In such a case, outer space itself will not be at 100-km altitude but will be kissing the surface of the earth thereby exposing everyone on earth to the harsh environment of space. The invisible gravither membrane, whose existence is beyond doubt, acts as the outer shell of spacecraft earth.

In this beautiful manner, nature has ingenuously not only provided the outer shell of the natural spacecrafts (i.e., the earth and all other celestial bodies) but also provided more than ample amount of space within these natural spacecrafts in which atmospheres are formed and life is flourishing. The atmospheres of planets, in reality, are the interiors of these natural spacecrafts and their surfaces act like the floors of spacecrafts.

Formation of gravitational fields around celestial bodies

We are now familiar with how mass contained in celestial bodies displaces the gravither existing in space due to which a planet's atmosphere and the elastic gravither membrane come into existence. Because of this displacement, the gravither present above this membrane, which is extremely sensitive to mechanical pressure, is also subjected to a spherical and symmetrical distortion because of which its density is not uniform. The density is highest just above the gravither membrane and decreases in accordance with the inverse square law as the distance from it increases. This presumption is based on the opinion of Newton and on the fact that if it was otherwise, then none of his equations would have worked so accurately. Due to the decreasing density of the gravither, at a particular altitude from the earth's surface, its density will be equal to the natural density of the gravither present in far out space and the influence of the earth (i.e., its Hill Sphere) will end at this point. In other words, the gravither up to this point is curved around the earth, and beyond this point, it is curved around the sun. The particular area where the gravitational fields of the earth and the sun meet or cancel out is known as the earth and sun Lagrange point. In case of the moon, it is the earth–moon Lagrange point.

The extent to which this distortion takes place is probably exponentially proportionate to the mass of a celestial body and will now act as the gravitational field of the body. This can be inferred from the following data: The moon has a mass of 7.35×10^{22} kg and its gravitational field extends up to about 60,000 km from its surface. The earths' gravitational field, whose mass is 81 times more than the moon, extends up to a distance of about 1.2 million kilometer. The mass of the sun is 333,000 times that of the earth and its gravitational field extends up to about a light year or more away from its surface.

From these observations, we can also infer that the extent of the gravitational field of a celestial body is solely dependent on its mass alone and is in no way affected by the mass of the sun. However, its distance from the sun is very important as the density of the gravither decreases with increase in the distance from the sun. Hence, the Hill Sphere of a body will be lesser when it is closer to the sun and will gradually increase as the distance from it increases.

From our discussions, it is clearly evident that without the presence of a massive heavenly body like the sun, earth, or moon, a gravitational field can never form or exist, as there is no displacement of the gravither present in space.

Mechanics of orbits

As we are now aware of how the gravitational fields around heavenly bodies come into existence and function, let us find out and understand how orbits actually take place.

It is universally accepted that in space, a weightless object in motion will always travel in a straight line when no external force is acting on it. The attractive force of earth's gravity acts as the external force and pulls it downwards toward the center of mass of the earth if it happens to be within its gravitational field. However, if it is moving with the required angular velocity, it will continue to fly in a circular or elliptical orbit, as a state of equilibrium arises between the two forces, that is, gravity and the centripetal force.

In this context also, the above explanation of orbit remains same except for one notable difference. As we now know that weight is an inherent property of matter, objects fall down toward the earth because the heaviness of matter contained in them gives rise to a downward force and not because of the pull of gravity.

There is another important aspect concerning orbits that need some clarification. Is the angular motion of orbiting bodies inertial or accelerating?

As rightly posited by Einstein, in his general theory of relativity, a celestial body in a circular or an elliptical orbit is not accelerating but just moving in its geodesic because although the earth's surface is curved in reality it behaves exactly like a flat surface. That is why if we fly a full circle around the earth, in an aircraft, at all times we will feel that we are traveling horizontally and in an upright manner. It almost feels as if we are stationary and the earth along with the sky is rotating.

Moreover, wherever or whenever there is acceleration, there has to be a force behind it and a force requires a constant source of energy. If the potential energy stored in orbiting bodies was being used to constantly change their direction, then their orbital speed would gradually decrease because of which they would eventually fall down and crash onto the earth.

The following thought experiment will prove that the orbiting motion of celestial bodies is inertial motion.

Thought experiment

Let us visualize a small stone tied to a string, which we are spinning in an anticlockwise direction in the vertical plane as illustrated in fig. 37.

Fig. 37

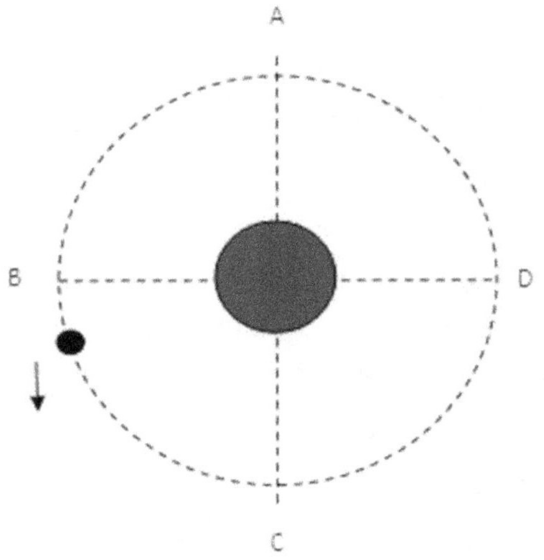

We will notice that as the stone moves downwards from point A to B and then C, it will accelerate and as it moves upwards from point C to D and again A, it will decelerate. In other words, its angular speed is constantly varying.

Our solar system is always illustrated as functioning in the horizontal plane. However, when we perceive it from the northern side of the ecliptic, it will appear in the vertical plane with the planets located at different altitudes. Let us assume that in fig. 37 the stone represents the earth and the central body is the sun and observe its motion around the sun. We will notice that as the earth moves from point A to B and then C, unlike as in the case of the stone, its angular velocity will not increase, and likewise, when it moves from C to D and then again A, its angular velocity will not decrease. Its movement along its orbit is exactly similar to the movement of aircrafts in earth's atmosphere. Throughout the earth's orbit, its angular velocity will always remain constant as long as its distance from the sun is unchanged. However, orbital velocity will increase when its distance from the sun decreases and vice versa due to the law of conservation of angular momentum.

This thought experiment proves Einstein's postulate that the orbital motion of celestial bodies is inertial and not accelerated. However, as mentioned earlier, the motion of freely falling bodies is definitely not inertial motion because they are in reality actually accelerating.

Acceleration in the absence of a gravitational force

Acceleration is a direct result of the potential energy objects possess by virtue of their distance above the earth's surface. They begin to fall down due to their heaviness and accelerate due to the potential energy, which at present is referred to as the gravitational potential energy.

Galileo Galilee proved Aristotle wrong by confirming with the help of simple experiments that all objects irrespective of mass fall down at the same rate, in the complete absence of friction. As the experiments prove beyond any doubt that an independent gravitational force does not exist, let us now find out why all objects, irrespective of mass, fall down with the same rate of acceleration.

According to the prevalent Newtonian theory of gravity, the attractive gravitational force emanating from the earth and acting on objects is always proportional to their mass due to which they all fall down with the same rate of acceleration.

However, according to NTG, the reason is exactly opposite. We now know that all objects irrespective of their mass comprise of atoms that consist of a nucleus containing matter, which is surrounded by the SM. The weight of matter present in the nucleus gives rise to a downward force representing gravity and the SM surrounding the nucleus gives rise to an upward force representing antigravity. This upward force, like before, is always proportionate to the mass of the objects. Because of this, if the mass of an object A is 100 times more than the mass of an object B, the upward force acting on it is also 100 times greater than that acting on object B. This is why all objects irrespective of mass fall down at exactly the same rate of acceleration in the absence of friction.

We should view all objects neither as individual elements or compounds nor as solids or liquids, but as a conglomeration of trillions of hydrogen atoms, which by the way is actually true because the basic building block of all the matter existing in the universe is the hydrogen atom.

Why does the rate of acceleration of freely falling bodies in space vary with altitude?

As we now know that weight is an inherent property of matter, the mystery behind why objects fall down and accelerate, even in the absence of gravity, is resolved. What we have to find out is why, even in the absence of a gravitational force, the rate of acceleration of a body in free fall, in outer space, increases as its distance from the earth's surface decreases and vice versa. This may have something to do with the gravitational field constituted by the gradient-density gravither surrounding the earth and all other celestial bodies definitely.

When a heavy object is dropped in earth's atmosphere from a height of say 1 km, it will fall down and accelerate at 9.81 m/sec^2 until it reaches a particular velocity after which it will fall at a steady speed known as the terminal velocity. This phenomenon is a direct result of the friction arising between the falling object and the atmospheric air. Because the gradient-density gravither is exactly similar to earth's atmosphere, one would expect the speed of freely falling objects in space also to decrease as they move closer to the earth. However, as the exact opposite is occurring, let us try to discover the reason behind it by observing the various phenomena that occur on earth where work is done by natural phenomena and not by the effects of a direct force.

The following are a few examples:

Examples

1) Osmosis - In this process, water moves from a lower density region (soil) to a higher density region (inside roots of plants and trees) through a semipermeable membrane.
2) Capillary action - Here a liquid rises up automatically in narrow vertical tubes as in tree trunks.
3) In earth's atmosphere, air moves from a higher to a lower density region or from a colder to a warmer region because of pressure difference.
4) Buoyancy - Here an object placed into a fluid is subject to an upwards thrust depending on the density of the fluid which reduces its apparent weight.

Of these examples, as only buoyancy seems to hold out some hope, let us conduct a few thought experiments on the subject.

Experiment 1

Let us visualize that we attach weights, with the help of strings, to a balloon filled with air and drop it in a deep lake until it just begins to sink. We will notice that as the contraption continuously descends the volume of the balloon will steadily decrease causing it to accelerate until it reaches the bottom.

Experiment 2

Let us conduct a similar experiment with a hydrogen-filled balloon to which an object of mass of 1 kg is attached. Fig. 38 illustrates the earth's atmosphere, in which the density of air is highest near the earth's surface and decreases exponentially with increase in altitude.

Fig. 38

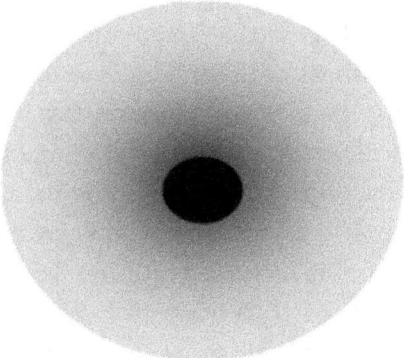

When the hydrogen-filled balloon, whose volume is sufficient to lift up the object to a considerable height, is released from the earth's surface, it will initially accelerate and then rise up steadily as the increasing volume of the balloon is offset by the decreasing atmospheric pressure until it finally comes to a standstill. Now, as before, let us visualize that additional weights are added to the balloon until it just begins to descend. We will notice that, just as in the earlier experiment, the contraption will steadily accelerate downwards because the increasing atmospheric pressure will compress the balloon due to which its volume will gradually decrease until it crashes on to the ground with maximum speed.

The results of these two experiments illustrate how buoyancy in specific cases can cause the rate of acceleration of freely falling objects to continuously increase when they are dropped in a gradient-density medium. As the laws of physics apply universally, let us find out if we can apply the same phenomenon to explain the relationship between altitude and rate of acceleration of freely falling bodies in outer space.

Fig. 36 illustrates earth's gradient-density gravitational field. We are now aware that when an object enters space it displaces an amount of gravither, which is proportionate to its mass. Let us visualize that a solid metallic sphere having a mass of about 1,000 kg enters space and reaches an altitude of 150 km from the earth's surface. It will displace an amount of gravither that is proportionate to its mass due to which a small gravitational field forms around it. The gravitational field, which is physical in nature, behaves as an integral part of the object and hence will move along with the object just as earth's atmospheric air and all other layers move along with it. Let us assume that the Hill Sphere of this gravitational field extends to about 10 m from the center of mass of the sphere. According to the NTG, just as the volume of an hydrogen-filled balloon increases with increasing altitude in earths' atmosphere, the extent of the gravitational field of the object will also increase when it moves to a higher altitude because the density of the gravither that makes up its gravitational field is greater than the density of the gravither existing at that increased altitude. For example, if the object is moved to at an altitude of 10,000 km from the earth's surface, then its gravitational field will be definitely more than 10 m. This is exactly similar to how the air pressure inside and outside a balloon in earth's atmosphere always tend to equalize. The gravitational fields of celestial bodies behave exactly as the hydrogen-filled balloons and cause the rate of acceleration of freely falling bodies in space to increase as they move closer to the earth's surface and vice versa. If we consider the moon, its Hill Sphere will increase if its distance from the earth increases and vice versa.

Because the laws of physics apply universally, the various physical phenomena (e.g., buoyancy and so on), which occur on the earth, also apply universally.

Lack of friction in space

A majority of the astronomers and scientists from the past, who believed that the earth and all the other celestial bodies moved through the ether just as a submarine moves through water, were confronted with a major problem; they could not elucidate why celestial bodies were not subjected to friction in space. Let us try to find out why.

Why does earth's rotation not give rise to friction?

Although the space surrounding earth is definitely filled with gravither, we now know that it exists only above 100-km altitude from its surface. If the gravither had been in direct contact with the earth's surface, friction however small would have slowed the earth's rotation and brought it to a standstill by now.

The earth's rotation does not give rise to friction because of two reasons.

1) The atmospheric air as a whole behaves as an integral part of the earth and rotates along with it. That is why aircrafts take the same amount of time to fly a particular distance irrespective of whether they are flying in the direction of the earths' rotation or in the opposite direction.
2) Most of the atmospheric air is present below 80-km altitude from the earth's surface, which is why spacecrafts re-entering the earth encounter friction, caused by air, only below this altitude. The space in between 80- and 100-km altitude is almost a pure vacuum as it is devoid of both a notable quantity of air and gravither.

The earth is suspended and rotating in this vacuum because of which friction is absent.

Why earth's motion through space does not give rise to friction?

There can be only one explanation for this, which is just as the atmospheric air, exosphere, thermosphere, and Van Allen belt, and so on behave as an integral part of the earth and move along with it in its journey through space, the gravitational field of the earth, which is also physical in nature and can be referred to as earths' celestial atmosphere, also behaves as an integral part of the earth and moves along with it as it orbits the sun. In other words, unlike a submarine that plows through water and is in direct contact with it, the earth is not in direct contact with the gravither but is suspended in the cocoon-like spherical structure formed by the spherically and symmetrically deformed gravither surrounding it. This cocoon as a whole, which represents the entire gravitational field, whose outer most regions are extremely low in density, moves through the equally low-density gravither that forms the celestial atmosphere of the sun and is curved around it. We are forced to come to this conclusion because not only do objects, like the moon, orbiting the earth move along with it, but also their orbits are also very stable. If the gravither existing within the gravitational field of the earth was not moving along with the earth, then there would have been tremendous turbulence all around the earth due to which forget stable orbits, orbits themselves would have been impossible as there would be no uniform physical connection between the earth and the moon including other bodies in space.

The gravitational field that forms around celestial bodies is exactly similar to the electromagnetic field except that it is physical in nature.

Let us visualize that we are holding lighted electrical bulb and are moving it randomly or in a circular motion. We will notice that the light emitted by the bulb will form a sort of halo around it and will always move along with it thus behaving as an integral part of it. Similarly, the gravitational field or the celestial atmosphere, which comes into existence because of the displacement of the gravither present around a celestial body and curves around it, behaves as an integral part of it and moves along with it in the course of its orbit.

The electromagnetic field always represents the energetic part of matter and is present in a condensed form as the nucleus of an atom. On the other hand, the gravitational field represents the physical aspect of matter and exists in a condensed state as the low-density elastic substance that surrounds an atom's nucleus.

Here, it would be appropriate to recall the observations made by the American physicist Albert Abraham Michelson, who himself was a firm believer in ether theories, after his experiment conducted in the year 1887, famously known as the Michelson–Morley experiment, failed to detect any ether drag.

"In any case we are driven to extraordinary conclusions and the choice lies between the three.

1) *The earth passes through the ether (or rather allows the ether to pass through its entire mass) without appreciable influence.*
2) *The length of all bodies is altered (equally?) by their motion through the ether.*
3) *The earth in its motion drags with it the ether even at distances of many thousands of kilometers from its surface."*

George Stokes, an English mathematician and physicist, proposed the following theory in the year 1845. According to his theory, the ether is completely dragged within and near matter, partially dragged at larger distances and stays at rest in free space.

Max Planck came up with another proposal in the year 1899, which implied that both conditions of Stokes theory (the ether movement must be nonrotating and its velocity near the earth's surface should be equal to the earth's velocity) could be met if it was granted that the ether was compressed by Boyle's law and was subject to gravitation. Around the earth, it is compressed like the atmosphere, the light velocity does not depend on the ether thickness.

Michelson's third possibility along with George Stokes and Max Planck's hypothesis strongly corroborates the new theory.

If we now recall the structure of an atom as illustrated in fig. 28, we will notice that it is exactly similar to that of the earth, which is surrounded by the gravither as illustrated in fig. 36. Just as the nucleus is subjected to a compressive force, the earth is also subjected to a compressive force.

The new theory also postulates that as the mass of a celestial body increases, the density and the thickness of the gravither membrane will increase proportionately due to which the intensity or strength of the mutual repulsion between the two will also increase proportionately. Consequently, the intensity of the compressive force acting on the body will increase proportionately as the mass of the celestial body increases and vice versa.

Reasons behind orbit decay experienced by satellites

Fig. 39

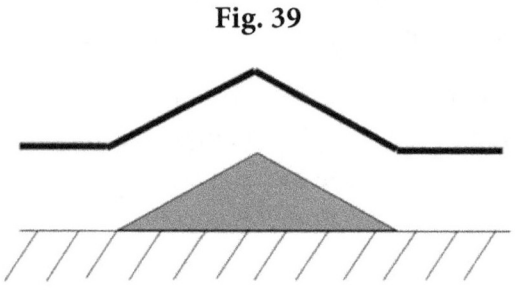

As the gravither membrane forms because of the mutual repulsion between the gravither present in space and the earths' surface, its unevenness is also transferred to the gravither membrane and the gravither behind it as illustrated in fig. 39. However, as the distance from the earth's surface increases, this unevenness gradually smoothens out until at a particular distance it will completely vanish. This is one of the main reasons why satellites are subjected to more orbit decay when closer to the earth than when they are far away from it. The second reason behind orbit decay is the gradient-density ether. Its density is highest when closer to celestial bodies and decreases in accordance with the inverse square law as the distance increases. Although satellites in space are not in direct contact with the gravither due to mutual repulsion between the two the displaced gravither, which forms a small gravitational field around it, is moving through the vast ocean of gravither similar to how water currents move in the oceans, giving rise to friction. That is why the satellites themselves will never heat up due to friction.

Why do earth and moon seem to attract each other?

Before we discuss about the earth and moon, let us perform a simple thought experiment to find out how the circular motion of a smaller body can cause a bigger body to wobble.

Thought experiment

Let us visualize a hollow sphere inside which a spherical ball is continuously moving, at such a speed that it always is in contact with the internal wall of the sphere. As the internal portion of the hollow sphere curves inwards, the center of mass of the ball will always be in line with the center of mass of the hollow sphere as long as it always remains in contact with it while revolving.

Now let us imagine that the hollow sphere is fixed to a slightly flexible rod, which in turn is fixed to a rigid wall in the horizontal direction. Exactly opposite to this rod (i.e., on the other side of the hollow sphere), a small pointed rod is also fixed as illustrated in fig. 40.

As the ball inside the hollow sphere rotates continuously in the clockwise direction, we will notice that the small pointed rod is also rotating in a small circle clearly indicating the fact that the hollow sphere itself is wobbling under the influence of the rotating ball. In a similar manner, the orbiting moon causes the earth to wobble because of which it rotates around earth-moon barycenter.

Fig. 40

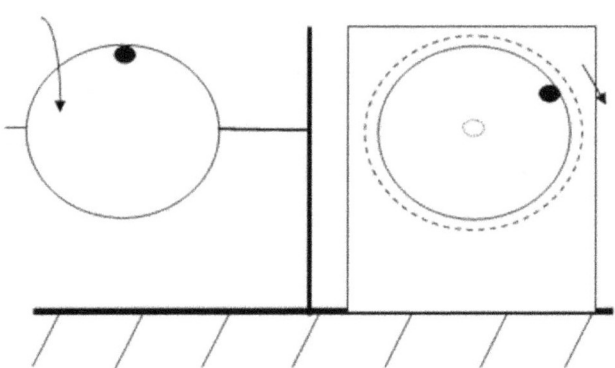

As we have discussed, like all other celestial bodies, the moon is also surrounded by a gravitational field, which moves along with it in its orbit around the earth. Although the diameter of the moon itself is only 3,000 odd kilometers along with its gravitational field, it measures around 120,000 km. The moon along with its huge gravitational field is tunneling through the gravither, which is curved around the earth and forms its gravitational field. We can compare the gravitational field of the earth to the hollow glass sphere that we discussed about in our thought experiment. The massive centrifugal force arising from the moons angular momentum causes the earth's gravitational field, which is invisible but physical like the air in earth's atmosphere, to shift slightly away in the outward direction. Because the gravitational field behaves as an integral part of the earth, the earth also moves slightly in the direction of the shift.

This gives rise to the illusion that the moon is exerting an attractive pull on the earth. As the direction of shift is constantly changing due to the moon's orbit, earth rotates around the common barycenter of the earth and moon as illustrated in fig. 41.

Fig. 41

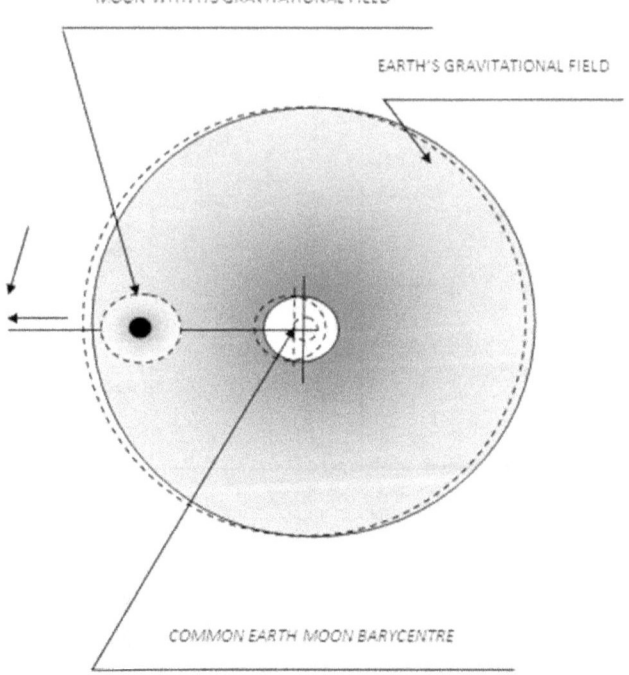

At the center is the earth and the shaded portion is the moon, which is orbiting the earth in an anticlockwise direction. The bigger dotted circle illustrates how the moon causes the gravitational field to shift. The smaller dotted circle is the earth, which is also shifting continuously.

The earth seems to attract the moon because the weight of the moon, which is considerably less at that distance from the earth, gives rise to a downward force. However, it continues to remain in orbit because it possesses the required angular velocity.

The net result of these interactions is that, even in the absence of any gravitational force emanating from either the center of the earth or the moon, they both seem to be attracting each other due to which the moon stays in its orbit, whereas the earth revolves around the earth–moon barycenter. This constant wobbling of the earth, indirectly caused by the moon, causes the ocean tides on earth and not any direct attractive force emanating from the moon.

What happens when an object from a celestial body enters space?

1) As soon as an object, irrespective of mass, enters space, it will behave like an independent celestial body and is immediately subjected to a compressive force arising from the mutual repulsion between the object and the gravither. This will also give rise to an atmosphere and a gravitational field, as explained earlier, which is proportionate to the object's mass.

2) Its entire weight will act toward its own center of mass and the gross weight acts toward the central body.

3) If it possesses the required orbital speed, it will go into a regular orbit. However, if its angular velocity is insufficient, it will fall down toward the central body.

4) If the body is hollow (e.g., a spacecraft), then the spacecraft will fall down toward the earth. However, all the unattached objects within it will begin to float as they are not in direct contact with the gravither.

What happens when an earthly body enters moon's atmosphere?

Let us consider the earth and moon and find out what actually happens when an object from earth enters moon's atmosphere and vice versa.

When a satellite from earth, which is in space, enters the moon's atmosphere, it will begin to experience weight for reasons mentioned earlier. However, the million-dollar question here is, how much will its weight be on the moon? Will it be one-sixth of its weight on earth as we have studied in schools or some other value? Its weight should be only slightly less when compared to its weight on the earth because we are now aware that two factors give rise to an objects weight. One is the heaviness of matter and the other is gravity, which is the compressive force exerted by the moon's gravither membrane. Its weight arising from the heaviness of matter contained in it will remain same because its density is unchanged. However, its weight arising from the moon's gravity will be slightly lesser than on earth because the moon's mass is 81 times lesser than earth's due to which the amount of gravither displaced is also proportionately lower due to which the density of moon's gravither membrane is comparatively lesser. As a result, the intensity of the mutual repulsion between the satellite and the moon's gravither membrane will also be proportionately lesser when compared to the intensity on the earth.

Similarly, as the density of a moon-rock on the earth's surface remains unchanged, its weight arising from the heaviness of matter will remain same as measured on the moon. However, its overall weight will increase slightly due to the increased compressive force exerted by earth's gravither membrane.

Why are stars not following Newton's laws of gravity?

Although the gravitational fields around stars obey the laws of gravity, the distant stars themselves seem to violate them. As the distance from the center of galaxies increase, the orbital speed of the stars should decrease. However, they seem to be moving with greater speed as their positions with respect to other stars, observed and recorded from hundreds of years, seem fixed. Moreover, the spiral shape of galaxies should have also disappeared long ago due to the winding effect. From these observations, it appears that the stars are not orbiting the center of galaxies and instead remain suspended in their respective places in space. However, in such a case, the stars should fall back toward the center of their respective galaxies, which is not happening. This phenomenon could be due to the following two reasons. The stars are so far away from the center of galaxies that the space there is isotropic due to which from their perspective all the directions are same and hence they remain floating. It is also possible that although the stars with respect to each other are motionless, the galaxies as a whole might be rotating. Only stars closer to the center of galaxies, where space is not isotropic, are orbiting rapidly according to the Newton's laws of gravity.

From earth size to a black hole

"All truths are easy to understand once they are discovered; the point is to discover them."

Galileo Galilee

From a single atom to earth's size, we have observed all the changes that took place in our imaginary body and how the effects of gravity came into existence.

The earth contains different elements, compounds, and other substances in various proportions and is a rocky planet.

From now on, let us presume that the mass of our imaginary earth-sized body continues to gain more mass in exact proportion to the various materials present on the earth and observe the results.

Formation of Jupiter-sized body

Let us assume that the mass of our imaginary body keeps increasing until it is equal to the mass of Jupiter. The immense pressure arising from the weight of the increased mass along with the compressive force exerted by the gravither membrane compresses the GM present in it to such a high level that most of the second matter present in the atoms near the core is squeezed out and escapes as gravither into space. As a result of this, the nuclei of atoms are densely packed together with absolutely no space in between just as in an atom's nucleus. Planetary scientists estimate that its core may be around 32,000 km in diameter and the pressure could be about 100 million atmospheres. Fig. 42 illustrates the density gradient of the Jupiter-sized body.

Fig. 42

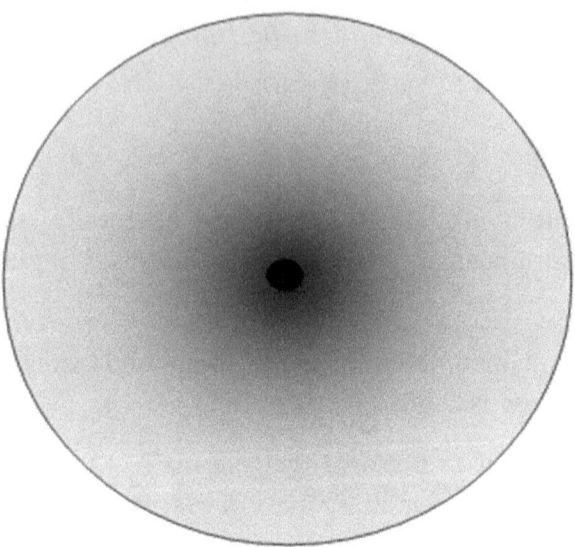

However, this does not mean that the density of the entire core is uniform. Logically speaking, even in this core, the density should be highest at the exact center of mass of the body and decrease gradually as we move away from it. Let us presume that the densest region of the core from where the entire second matter has been squeezed out and which resembles an atom's nucleus is about a kilometer or two in diameter. If we now view this dense core from the perspective of the contemporary atomic theory, nuclear fusion has taken place. However, there is a huge difference between this fusion and the fusion process according to the new atomic theory. According to the new theory, during nuclear fusion, the SM present in atoms is not squeezed out of the system but fuse together to form a new element and the electrons remain in the newly formed atom. However, in the case of Jupiter's core, no new element is formed as the entire SM has been squeezed out from the core. In fact, its core is exactly similar to the nucleus of an atom of a heavy element like uranium. Both contain only protons and neutrons in a densely packed state. This means that just as the uranium atom's nucleus is prone to nuclear fission so is Jupiter's core. In case of the uranium atom, the second matter surrounding its nucleus prevents it from exploding. Jupiter's explosive core does not explode because it is surrounded by highly dense

GM and is also subjected to a compressive force equal to around 100 million atmospheres. As far as the electrons are concerned, two things could be happening to them. One they could disassociate from the atoms, rise up toward the surface, and exist as free electrons because of which the earth and other huge celestial bodies possess a net negative charge. Second, they could combine with the protons to form neutrons or a combination of the two is also possible.

Here, let us recall the chapter on atoms in which we discussed the reason why elements with high atomic numbers become radioactive on the earth.

As the quantity of SM in an atom is reduced, the intensity of the compressive force acting on the nucleus decreases due to which the element becomes radioactive. If an element was already radioactive on the earth, then its half-life is further reduced on Jupiter. For the same reason, elements that were normally not radioactive on the earth would become radioactive if they happened to be in or near Jupiter's core.

For reasons mentioned earlier, any element present in the core will lose its identity with its nucleons becoming a part of the super dense core. However, elements that are further away from the core and that have lost only part of their SM will retain their elementary form, but they will be more radioactive than they had been on the earth. However, as on the earth, elements will become radioactive in order of their atomic numbers. For example, lead will become radioactive before copper or iron and so on.

If we now consider Jupiter in this new perspective, the core is undergoing nuclear fission on a massive scale and there is a reduction in the half-life of substances, which were already radioactive, due to which they decay at a faster rate than they were decaying on the earth. Additionally, there is substantial increase in the number of elements that have now become radioactive due to which the following changes occur in Jupiter.

1) The temperature of the massive nucleus-like dense core increases phenomenally. Its temperature is estimated to be around 20,000°C. This heat is gradually transferred by convection to the surface due to which the overall temperature of Jupiter increases and it begins to radiate more energy than it is receiving from the sun.

2) Due to the extreme heat, a portion of the rocky matter present in the body is slowly transformed into a gaseous state. These gases gradually move toward the body's surface and over a period of time will cover the entire surface of Jupiter.

That is why all huge celestial bodies always appear to be gaseous while smaller bodies are always rocky.

If we now consider the four Jovian planets in our solar system, they may actually contain molten rocky matter that is hidden under the huge quantity of gases present in their atmospheres.

Conclusions

1) Nuclear fission not only depends on the atomic number of an element but also on its atomic radius. Hence, as the average density of a celestial body increases due to increase in mass, more elements will become radioactive and the half-life of elements, which were already radioactive, will decrease proportionately.
2) All celestial bodies, which can be visualized as ultra huge atoms, also behave like atoms and as their mass increases beyond a certain limit, a dense nucleus-like core will form at their center of mass because of which nuclear fission commences. The diameter of this super dense core will increase as the mass of celestial bodies increase and vice versa.

Formation of the sun and stars

As the mass of our imaginary body continues to increase, the diameter of the super dense core will also increase leading to enhanced levels of nuclear fission. At the same time, an increasing number of elements tend to become radioactive. Because of these two reasons, the average temperature of the body will also increase proportionately.

When its mass approaches the mass of our sun, the size of its super dense core, which is constantly subject to nuclear fission, might have increased to thousands of kilometers in diameter. The pressure here is estimated to be around 200 billion atmospheres and temperatures around 16 million kelvin. Hence, most of the matter present in this region will be in plasma state, which is considered as the fourth state of matter. This huge body has now transformed into a star.

It is now clear that even a body comprising of entirely rocky matter will transform into a purely gaseous state when its mass increases substantially, until ultimately it becomes a star. To transform into a star, mass is the only critical factor involved. The nature of elements contained in the body is irrelevant as they all lose their elementary form as they are transformed into a huge mass of plasma inside the dense core. Just as death is a great equalizer among humans, so are the cores of huge celestial bodies with respect to the elements.

Here, let us discuss a little about the gravither membrane. The earth's gravither membrane is just about 20 km in thickness and completely invisible as the mass of earth is extremely low when compared to the sun. However, as the sun's mass is 330,000 times that of earth, its gravither membrane should be visible as its thickness and density would also be proportionately higher. As we have discussed earlier, the gravither membrane clearly separates space and a heavenly body's atmosphere. In the empty space existing in between the membrane and the body's surface, the atmospheric gases can be found in the lower areas, whereas the upper portion is a pure vacuum devoid of even gravither. Let us try to locate the gravither membrane of our sun. The sun has only three atmospheric layers namely photosphere, chromosphere, and the corona. The photosphere measuring 500 km represents the sun's inner atmosphere, the chromosphere, which is about 2,000 km in thickness, represents the gravither membrane, and the corona represents the beginning of its gravitational field.

Formation of black holes

"Facts which at first seem impossible will, even on scant explanation, drop the cloak that has hidden them and stand forth in naked and simple beauty."

Galileo Galilee

As the mass of our imaginary star keeps increasing, its brightness and temperature will also increase proportionately. Simultaneously, the thickness and density of the gravither membrane will also be increasing proportionately. Because of this, after a certain point, the brightness of the star will slowly appear to diminish, even as the star grows bigger and bigger. Let us assume that its mass further increases until it is equal to a few million or billion solar masses and observe the changes that have occurred.

At this stage, all the SM present in the atoms of this huge body would have escaped and merged with the gravither membrane surrounding it from outside. What remains is a highly dense structure of pure matter. This state is exactly similar to the how matter exists in an atom's nucleus on earth, but with two exceptional differences, which are as follows:

1) Its temperature will probably measure in millions of degrees Celsius.
2) All the protons, neutrons, and electrons would have merged and will exist as a single homogenous structure of pure matter.

This huge and extremely explosive object is a super massive black hole. Just like the SM in an atom, the highly dense, thick, and repulsive gravither membrane tightly binds and holds together this explosive mass of mind-boggling energy. This gravither membrane now contains all the SM that was earlier present in the atoms of the newly formed black hole, as well as the gravither that was present in outer or galactic space. This dense gravither membrane could be a few 100,0000 km in thickness. That is why although the temperature inside the black hole is extremely high, the light and heat energy are not visible or felt from outside the gravither membrane. The inner portion of the membrane is intensely repelling the matter inside thereby exerting a crushing force on the extremely hot and dense molten matter, whereas the outer portion is exposed to the extremely cold and harsh environment of galactic space.

This is why the instruments and other methods used to measure the temperature of a black hole always indicate it as an extremely cold object.

The gravitational field of the black hole will now span across hundreds or thousands of light years from the outer surface of the gravither membrane. All the stars and our solar systems are actually orbiting the black hole at the center of the galaxy and within its gravitational well. Without the presence of a huge object, there can be no gravitational field and without gravitational fields, orbits are not possible. In the absence of a black hole, heavenly bodies like the stars and planets will drift away into space, with nothing to control or keep them together.

We can now describe a black hole as a body whose mass is so great that it has completely displaced all the SM surrounding the nuclei of the atoms contained in it. We can also describe it as an ultra huge atom floating in the vast expanse of galactic space, with billions of stars and solar systems orbiting around it.

We have now visualized how a single atom grows into the size of the earth, the sun, and a super massive black hole and what happens inside and outside the object, as it grows in size. Let us summarize all that we know of the black hole in this new perspective.

Summary

1) A black hole is a massive, spherical object made up of pure and extreme dense matter, whose temperature may measure billions of degrees centigrade.
2) The highly dense and repulsive gravither membrane, measuring a few 100,0000 km in thickness, binds and holds together this extremely dense, hot, and explosive matter. This membrane by virtue of its density and thickness prevents the heat and light, from escaping through it. As no light is visible outside the object, it appears dark and hence invisible as well as cold.
3) The gravither membrane is surrounded by a strong gravitational field that extends thousands of light years away from it.
4) No hole exists in its center and it does not gobble up other stars or planets, as is the popular belief. However, as the density of the gravither is extremely high in the close surroundings of the black hole, the objects within this range have to orbit at extremely high speeds to escape from crashing onto the black hole.

5) Super massive or primordial black holes must have formed at the beginning of the universe because of the huge mass present at that time. The smaller or stellar black holes form after an aged star having sufficient mass has exploded in an event called supernova.

Because of the reasons stated above, it would be highly appropriate if we henceforth refer to these super massive black objects as black wholes instead of black holes.

Formation of neutron stars and stellar black holes

"No great discovery was ever made without making a bold guess."

Sir Isaac Newton

Neutron stars and stellar black holes

Let us now find out how neutron stars and stellar black holes come into existence. However, before we proceed any further, let us observe what happens when a huge explosion occurs in the earth's atmosphere. After an explosion, which occurs in a fraction of a second, the energy released and the remaining matter if any will move away from the point of explosion in all directions, never to come back and that is the end of it.

Neutron stars and stellar black holes definitely form after an aged star has exploded which is known as a supernova. However, there is a difference between an explosion that occurs on the surface of a planet and a supernova.

Although the gravither membrane surrounding a star is very dense and thick, it is still elastic in nature and hence can be visualized as an extremely thick-skinned rubber balloon. As the explosion occurs, the gravither membrane stretches away to a distance, which depends on the intensity of the explosion.

We have to also remember that unlike in a balloon on earth, the thickness and density of the membrane will increase as it is pushed further and further back. Even though the membrane will easily allow some of the heat and light energy to escape through it, when it comes to matter, only small particles with high kinetic energy will be able to escape through it, that too at the beginning of the explosion. Hence, the quantity of GM that totally escapes from the system is minimal and most of the energy released by the explosion remains confined within the system and also in the gravither membrane as potential energy.

As the explosion subsides, the gravither membrane will move inwards with high speed giving rise to a crushing force. As it moves in closer toward the center of mass of the exploded star, the compressive pressure generated is millions of times greater than what existed before the explosion. This compresses the atoms in the star to such a great extent that all the second matter present in the atoms escapes out of the newly formed body leaving the dense nuclei behind. The density of matter in the star will now be so high that the star becomes either a neutron star or a stellar black hole, measuring only a few kilometers in diameter.

We have to think of explosions of huge objects in space, as always happening inside a thick-skinned balloon. This is because the thickness and the density of the elastic gravither membranes that surround heavenly bodies are directly proportional to the mass of the heavenly bodies.

Gravity waves

Let us now observe the happenings outside the gravither membrane during and after the explosion.

As soon as the explosion pushes the gravither membrane outwards, the extent of the gravitational field surrounding it also increases exponentially, probably at the speed of light. For example, if the gravity well of the sun extended till 1 light year away from it before the explosion, at the peak of the explosion, it would probably increase to one and a half or even 2 light years. However, as the explosion subsides, the gravither membrane returns to surround the newly formed dense body from a close distance. The extent of the gravitational field will now be only slightly lesser than before because although the diameter of the newly formed neutron star or a black hole is now only a few kilometer across, its mass will be only slightly lesser than what it was before the explosion. Only a small percent of its original mass is lost during the explosion.

The explosion also gives rise to gravity waves, which are mechanical ripples in the all-pervasive ocean of gravither. However, unlike light and electromagnetic waves, these waves will not travel very far because the elastic nature of gravither will dampen and absorb these shock waves. This is why the massive and extremely sensitive instruments constructed to detect gravity waves have not yet detected anything.

Summary

From the above observations, it is very clear that gravity is not an attractive independent force by itself but is an illusion caused by various factors. When in a planet's atmosphere, a body falls down and accelerates mainly because of the heaviness of the matter contained in it and the potential energy stored in it. The intense repulsion between matter and SM contained in atoms gives rise to the strong nuclear force and the less intense repulsion between celestial bodies and the gravither present in space gives rise to the gravitational effects that we have observed. Its long range of action, which although is not infinity, is possible because the extremely low density and elastic gravither pervades the entire continuum of space that makes up our universe.

Another interesting and important fact that has emerged from our discussions on gravity is that, gravity or the effects of gravity come into existence only when there is a heavenly body. Without a heavenly body, the effects of gravity are nonexistent because then there will be no displacement of the gravither present in space, and without displacement of gravither, the gravitational field or well will never come into existence.

If this is true, which primary or central body are all the stars and solar systems in our galaxy and other galaxies orbiting? The answer is now obvious. It is the super massive black holes present in the center of all galaxies. All the stars, solar systems, satellites, and individual bodies are actually moving in the gravity well created by super massive black holes. If the constantly orbiting heavenly bodies within a galaxy were to come to a standstill, then all planets will fall onto their suns and the stars in turn will fall back onto the black holes.

From the above observations, we can say that galaxies form around black holes, and hence no galaxy can exist independent of a black hole at its center.

Fig. 43 illustrates the structure of a galaxy.

Fig. 43

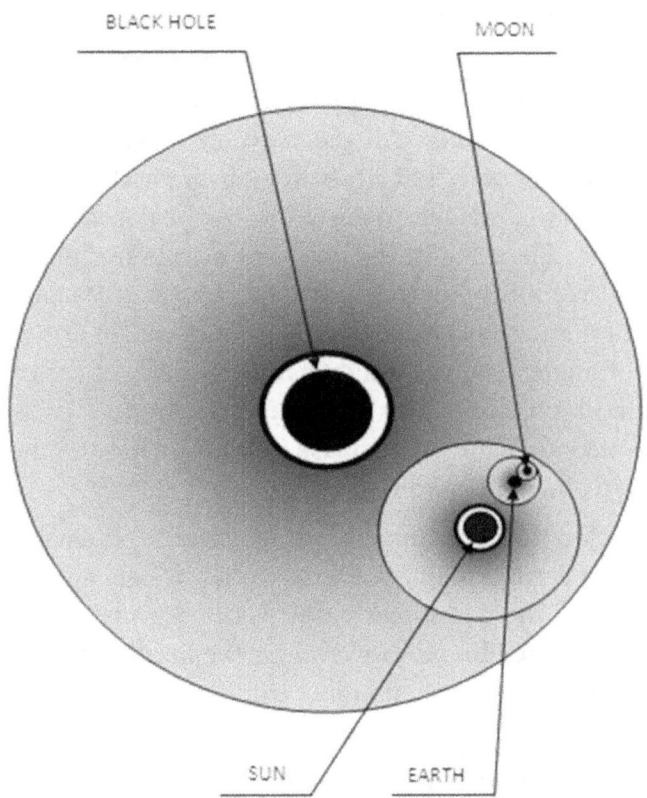

The central body is a black hole along with its gravitational field in which the sun, earth, and moon are also illustrated along with their gravitational fields.

Summary of gravity

Let us recapitulate our discourse on gravity.

There are three distinct features of gravity namely inner gravity, central gravity represented by the gravither membrane, and outer gravity.

Inner gravity

1) In the inner gravity region, the primary force that holds matter together at the atomic and molecular levels and subsequently huge celestial bodies are the intermolecular forces.

2) Mass and weight are the same and always remain constant except for the fact that external forces like buoyancy and so on influence the apparent weight of a body. All the movable objects on the earth or any other heavenly body remain firmly on the body because of their own weight, that is, weight arising from the heaviness of matter. The compressive force exerted by the gravither membrane, present high up in the sky, also contributes to the weight of objects. Weight of objects on a planet's surface will be equal to the sum of weight arising from the heaviness of matter, and the weight gained because of the compressive force exerted by the gravither membrane. The ratio of weight gained due to the second factor will increase proportionately, as the mass of a heavenly body increases and vice versa.

3) The acceleration that all falling objects experience on smaller heavenly bodies like our earth is mainly because of their weight and the potential energy stored in them and partly due to the downward force exerted by the gravither membrane. However, as the mass of a heavenly body increases, the gravither membrane will repel falling objects more intensely making them to accelerate at a faster rate and vice versa.

4) As the mass of a celestial body increases, its average density and weight will increase because the second matter that is squeezed out of the body escapes into space as gravither. Once the mass exceeds a certain limit, the cores of these bodies become radioactive due to nuclear fission.

5) Even in the absence of gravity, celestial bodies are spherical in shape as most of them must have been in a molten state at the time of their formation.

Central gravity (gravither membrane)

1) The gravither membrane clearly separates the inner gravity (atmospheres) and outer gravity (space) regions and protects celestial bodies from the harsh environment of outer space. It functions as the outer shell of a spacecraft.

2) It prevents the gases in the atmosphere from escaping into outer space. Atmospheric pressure is the result of the mutual repulsion between air and the gravither membrane.

3) The uniform compressive force exerted by the gravither membrane also plays a vital role in the spherical formation of celestial bodies, especially huge ones.

Outer gravity (gravitational field)

1) The outer gravity consists of static gravither whose density decreases in accordance with the inverse square law as the distance from the central body increases. This gradient density gravither serves the purpose of extending the influence of the primary body, by practically creating a three-dimensional well into which objects coming close enough will fall. Because of this, we always think of gravity as an attractive force, which is emanating from celestial bodies.

2) The outer gravity or the gravity field acts as an integral part of the primary body and together they behave as a single entity. The gravitational field constituted by the gradient-density gravither will travel along with the primary body as it orbits the sun. However, it will not rotate along with the primary body as there is no direct contact between the gravither membrane and a celestial body's surface.

3) Although all objects in this region or anywhere else in space are always in full possession of their weight, they appear to be weightless because the entire weight always acts only toward the center of mass of the objects and the gross weight in the direction of the primary body in whose gravitational field they are orbiting. Their own gravitational field, which is physical in nature, behaves exactly like hydrogen-filled balloons and expands when the distance from the primary body increases and vice

versa. This is why the rate of acceleration of a body in free fall in space increases with decrease in distance from the primary body and vice versa.

4) The gravitational field around a heavenly body has a definite boundary, although not as sharply defined as the gravither membrane. By correlating the mass of celestial bodies and their observed gravitational range, we can safely infer that the extent of this gravitational field is probably exponentially proportional to their mass. The presence of a second body near it or the sun does not in any way influence the extent of this gravitational field. This is the reason why orbits are not possible according to the Hill Sphere or Hill Radius of a primary body. Once a secondary body goes outside this boundary, it will cease to orbit the primary body. It would have now entered the gravity well of a different body.

5) The gravither membrane along with the gradient-density gravither, which surrounds heavenly bodies in outer gravity region, actually acts as huge shock absorbers. If a massive celestial body, like a dying star, were to explode (supernovae) the gravither membrane, which stretches outwards, will temporarily absorb the expansive force created by the above event. Once the explosion subsides, the stored energy and matter shoots back toward the source of the explosion because of the massive compressive force created by the contracting gravither membrane to form a neutron star or a stellar black hole. This is why, the vibrations arising from the explosion (gravitational waves) do not continually flow outwards for long distances in space, like radio waves or the waves that form on water. They are absorbed by the elastic gravither pervading the continuum of space.

6) Another interesting phenomenon caused by the gradient-density gravither is the bending of light. When light from stars passes close to huge heavenly bodies like the sun, it causes light to bend because of refraction. Newton was the first to predict the occurrence of such phenomena, but strangely, this was somehow forgotten in the long passage of time after his death. Henry Cavendish in 1784 and Johann Georg Von Soldner in 1804 had also pointed out that Newtonian gravity predicts that starlight will bend around a massive celestial object.

Later on Einstein also predicted that the same phenomenon would occur and it was this that made him instantly world famous.

As the entire theory is based on the existence of gravither in space in a dilute form and in a condensed state, in GM it is of utmost importance to try to establish its various properties.

Properties of gravither

Although GM can exist in four different states and also as different elements and compounds, the gravither present in outer space has only one state of existence and this state has the properties of solids, liquids, and gases combined into one. Taking into account the various observations that we have made, we may not be wrong if we posit that gravither possesses the following properties:

1) It is comprised of particles similar to the photons, but unlike them, it does not possess any electrical or magnetic charge. Hence, instruments used for measuring electromagnetic forces cannot directly detect it.

2) In empty space, its density is extremely low and hence does not affect the speed of light. However, in the immediate vicinity of huge celestial bodies, its density increases due to displacement caused by them causing light passing close to such bodies to bend because of refraction.

3) It is physical in nature and just like the atmospheric air; it possesses elastic properties and hence can be compressed. However, unlike the atoms and molecules that constitute air, its particles do not possess any sort of energy.

4) By itself, it exists in a static, that is, motionless state, with respect to the black holes existing in the center of galaxies. However, a part of it will act as an integral part of any celestial body that has displaced it, and hence, it will traverse along with the body in the form of its gravitational field.

5) It is absolutely noncombustible and can withstand extremely high degrees of heat and cold. This can be inferred from two things. The extremely cold temperatures in space and the extreme heat in and around the sun do not seem to affect it.

6) By itself, as present in outer space, it is a nonconductor of electricity and heat.

7) The gravither present in space and the GM present in celestial bodies mutually repel each other. The intensity of this repulsion increases when the mass of a celestial body increases and vice versa.

8) In a condensed state, it constitutes the physical part of GM (i.e., second matter), whereas the nuclei of atoms, which are condensed electromagnetic waves, constitute the energetic part of GM.

9) When gravither is in the form of second matter, gravitationally it always acts opposite to the direction of matter. For example, if we consider atoms on earth, the weight of matter present in the nuclei of atoms always act downwards (i.e., toward the center of mass of the earth). However the SM surrounding the nuclei of atoms, which is concentrated gravither, will always act in the upward direction (i.e., toward space), which is an ocean of gravither.

Elements whose second matter is closest in similarity to gravither are hydrogen and helium. That is why they are the lightest elements, which tend to rise upwards and their boiling point is so low.

Right from the elementary particles, which constitute the nucleus of an atom to a super massive black hole, matter is always surrounded by this dilute and elastic gravither.

List of eminent personalities connected to NTG

As we have completed our study of gravity, let us recall the eminent philosophers, scientists, and personalities whose theories and postulates have been utilized in compiling this NTG.

The story of gravity begins with Galileo Galilee who was the first to discover that all objects irrespective of their composition and mass accelerate at the same rate during free fall in the absence of friction. Johannes Kepler predicted that the orbital paths of planets are actually ellipses and not circular. Next, Sir Isaac Newton discovered and wrote the universal laws of gravitation. Since his later theories of gravity failed to gain acceptance, they were not taught in educational institutions. In the long passage of time, they were completely forgotten by academicians and remained hidden in archives.

Hence, when Albert Einstein first proposed that gravity was not a true force and bodies in space followed a curved path as the space around huge bodies is curved due to space-time, it created a great sensation. Even though it may not actually be due to space-time, NTG has explained the reason for the curvature of space around heavenly bodies, which is a repetition of Einstein's later views expressed in his 1920 lecture. The secondary bodies follow the curved path or geodesics, exactly as explained in GR, but again for a different reason.

NTG is based entirely on the discoveries and theories of the following people:

1) A major portion of NTG is adapted from Sir Isaac Newton's mechanical ether theory of gravity published in the second edition of Optics in the year 1717.

2) In 1918, Sir Ernest Rutherford made a sensational discovery regarding the structure of an atom. He discovered that an atom consists of a nucleus, which is extremely dense and small and is surrounded by a vast amount of empty space that constituted almost 99.9% of an atom's volume. The idea of NTG has in fact originated from this empty space.

3) Paul Dirac's prediction that positron (antimatter and ether) should have existed in the universe forms the backbone of NTG.

4) NTG also depends heavily on the views of Rene Descartes, Sir Isaac Newton, Maxwell, Nikola Tesla, and a host of other eminent persons according to whom not only was space pervaded with elastic ether but also the ordinary low-density matter found in all objects was in fact a condensed form of that very same ether.

5) Leonhard Euler's hypotheses, which postulates that the ether present around celestial bodies is in a compressed state and hence exerts a compressive force on them, form an important part of NTG.

6) Another interesting person that we should remember here is Archimedes, who discovered and wrote laws on buoyancy known as Archimedes principle because gravity on a heavenly body as well as in

space works on the principle of buoyancy. Buoyancy requires only two things: an object and a medium.

The weight of matter present in a heavenly body varies mainly because of buoyancy.

Similarly, when we consider a galaxy, the gravither present in space becomes the medium and all the heavenly bodies are the objects.

It is indeed strange and fascinating to know that the universe functions almost entirely because of a law discovered and written in the year 212 BC by Archimedes.

7) A philosophical twist, regarding the dualistic nature of the universe, was provided by the ancient philosophical texts, which propounded that the universe is a product of two different or opposite energies.

However, we should consider Sir Isaac Newton as the father of gravity since he was the first person to write in detail about it. Though gravity is not an independent attractive force, he was the first to speak of the gradient-density ether. Most of his laws of universal gravitation are valid even today and will be forever because they are based on physical observations. His extraordinary and precise observational skills and mathematical capabilities, coupled with Johannes Kepler's laws of planetary motion, made him aware that heavenly bodies behave as if they attract one another and resulted in him writing the universal laws of gravitation and the various equations associated with it. Though heavenly bodies do not attract each other directly as we have envisaged, like say two magnets with opposite poles, the overall effect is exactly similar.

Experiments to test the veracity of the new theory

The results of the experiments mentioned at the end of misconception number one will clearly establish whether the new theory holds any water. Other than them, the following experiments will also help in establishing the veracity of the new theory.

1) **Experiment to check for antigravity properties**

    The medium present between 100- and 120-km altitudes from the earth's surface should be collected in a metallic cylinder, whose empty weight is already measured on the earth, and then brought back to earth. If the new theory is correct, its weight should now indicate a small decrease, which will prove that not only is space filled with a medium but it has antigravity properties also.

2) **Experiment to check whether it can be transformed into a physical entity**

   When a sufficient quantity of it is collected and subjected to continuous compression, it should assume a physical form akin to a tenacious gel or rubber-like substance.

3) If a rocket is launched vertically into space, it will be noticed that irrespective of the altitude, it is at an appropriate and constant thrust depending on its total mass, minus mass of its fuel, is always required to keep it ascending. However, according to the current theories, the thrust required will constantly decrease with increase in altitude as the weight of the rocket is also constantly decreasing.

4) **Comparing earth with sun**

   Let us now test the new theory by observing the structure of our sun as illustrated in fig. 44 as its mass is 333,000 times that of earth and try to find any evidence that can substantiate the postulates of the NTG.

Fig. 44

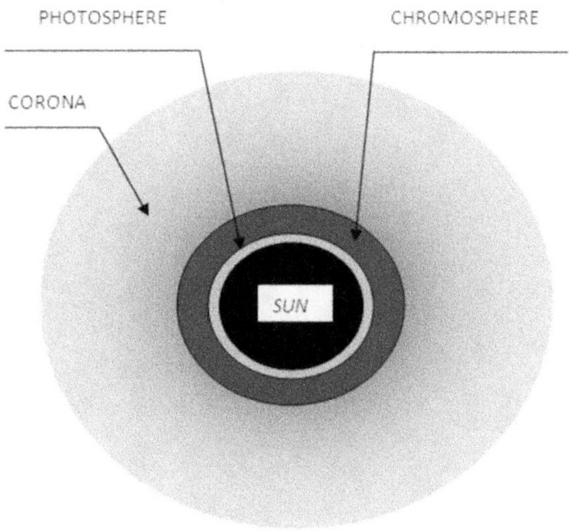

The sun's atmosphere consists of three layers. The first layer is the photosphere, which extends up to 500 km from its surface. Above the photosphere is the chromosphere that extends up to 2,000 km and above it is the corona. Isn't this figure strikingly similar to fig. 21, which illustrates earth's inner atmosphere, the elastic gravither membrane, and its gravitational field, and also fig. 28, which shows the structure of an atom? In the case of the earth, we arrived at fig. 21 by observing the different physical phenomena that occur in the various regions surrounding it, and in case of the atom, the quest for a logical explanation for the extremely short range of the strong nuclear force helped us in developing fig. 28. However, in the case of the sun, we are able to directly observe the sun's atmospheric structure. If we apply the new theory to the sun, then the photosphere will represent the sun's inner or terrestrial atmosphere, the chromosphere will represent the elastic gravither membrane, and the corona will represent the inner most regions of its gravitational field, which of course extends way beyond the corona.

If we go according to the contemporary theory according to which space is just an empty vacuum and the gravitational force emanating from its center holds the sun together, then the sun should have appeared as a blazing ball of flames unlike the smooth outer surface that is presently visible around it. The photosphere and the chromospheres have no business to be present around the sun. This is why astronomers and astrophysicists find it hard to explain their presence as well as their composition and purpose.

6
Misconception number 6

> *"To kill an error is as good a service as, and sometimes even better than, the establishing of a new truth."*
>
> Charles Darwin

Antimatter is missing from our universe

In 1928, Mr. Paul Dirac, a British theoretical physicist, predicted with the help of certain calculations (arcane mathematics) that positrons, which are the opposite of electrons, should and must exist in the universe. A few years later in August 2, 1932, Mr. Carl D. Anderson discovered the positron thus proving Paul Dirac's prediction right. This subsequently came to be known as antimatter, which is actually a mirror image or copy of matter but with opposite electrical charges. If we consider an atom of hydrogen of normal matter, its proton is positively charged and its electron is negatively charged. However, the proton possesses a negative charge in the case of antimatter, whereas the positron (opposite of electron) possesses a positive charge.

Scientists are 100% sure due to various calculations and theories that at the beginning of the universe, nature created matter and antimatter in equal quantities. Now, as only matter is visible and found in the entire universe, everybody is deeply perplexed as to where this huge quantity of antimatter has disappeared or vanished without a trace.

Experiments conducted at the LHC at CERN have been successful in actually producing antimatter atoms of hydrogen and preserving them for a period of 1,000 seconds by electromagnetic confinement at very low temperatures, thus confirming that it should have been present in the universe.

However, even at this low temperature, the antimatter atoms of hydrogen eventually disintegrate within a fraction of a second and vanish.

Let us now study the different views or theories regarding matter and antimatter that are prevalent today.

Some scientists believe that if brought together, matter and antimatter would instantly annihilate each other completely into energy. A question that arises from such an assumption is why nature would produce matter and antimatter if the purpose were to let them annihilate each other.

A few scientists are of the view that matter and antimatter will repel each other very strongly because although their elementary particles possess opposite charges and might annihilate each other, the atoms themselves as a whole are neutrally charged.

The classical isodual theory of antimatter proposed by the R.M. Santilli Foundation predicts the existence of antigravity for antimatter in the field of matter and vice versa. In this new perspective, if we assume for a moment that we are in possession of a small quantity of antimatter and place it on a table, instead of staying put on the table, it should rise upwards because gravity attracts objects made of matter toward the earth and objects made of antimatter will be repelled. In other words, matter represents gravity and antimatter represents antigravity.

However, as we have not yet discovered physical objects made of antimatter either in a solid or liquid state, there is no proof to support either theory.

Another interesting statement attributed to Paul Dirac is as follows, *"It is better to say that a vacuum or nothing is the combination of matter, and antimatter—particles and antiparticles. Their density is tremendous, but we cannot perceive any of them because their observable effects entirely cancel each other out."* This clearly shows that he believed that space might be full of matter and antimatter in an invisible state.

During the early and mid-90s, scientists of various countries tried to produce materials with antigravity properties, but were unsuccessful in their attempts. This might be probably because, only antimatter has the properties of antigravity and this substance of course seems to be missing from our universe.

Whatever it may be, as matter is existing, we must presume that annihilation has not occurred. In that case, we have to accept one of the following two possibilities.

1) If matter and antimatter were really to annihilate each other, which seem to have not occurred, then a different or parallel universe made up of only antimatter should exist. Many scientists and theorists including Paul Dirac have already proposed and supported such a possibility.

2) If matter and antimatter do not annihilate but will only repel each other, then there is a strong possibility that antimatter could be present in this very universe.

The million-dollar question is, where exactly could this antimatter be present or hiding?

Whatever may have happened to it, antimatter definitely does not exist in solid, liquid, or a gaseous state because, if it did exist in any of the above states, scientists would have already found it by now.

Before proceeding, let us study the views (about the universe) expressed in some of the ancient philosophical texts, written and propounded by highly competent philosophers, thousands of years ago.

There is a strong belief in some ancient Asian cultures and civilizations like the Indians, Chinese, and probably others that most of the opposites in nature exist side by side. This is known as the theory of dualism. According to these theories without the one, the other cannot exist. Examples include God and devil, man and women, good and bad, light and darkness, love and hate, sunrise and sunset, hot and cold, positive and negative, wise and foolish, life and death; the list is endless. It seems to be nature's way of keeping everything in equilibrium.

Even some Greek philosophers believed in the theory of dualism. Heraclitus discovered a new theory, which is as follows: *"the real world consists in a balanced adjustment of opposing tendencies. Behind the strife* between opposites, according to measures, there lies a hidden harmony or atonement which is the world."

> **"Opposites come together and from what is different arises the fairest harmony."**
>
> Heraclitus

Here I would request my readers not to get confused between religion and philosophy because religion is man-made and depends on one's personal choice and belief. However, philosophy encompasses everything, right from bacteria to plants and animals, to human life, intelligence and consciousness. In fact, philosophy is universal as it connects everything including the human mind with the entire universe.

The following is an excerpt from the book, *Summer Showers* in Brindavan 1993.

"The Samkhya system of Indian Philosophy propounded that the Universe is a harmonious product of paradoxical dualities with inherent harmony of the Universe.

According to this system, 'Paramatma' [Purusha or God] is latent in 'Prakruti' [nature] as oil is latent in seeds, as fire in wood and as fragrance in flowers. This system holds that creation cannot be made of one entity alone and that it is the unison of two entities, viz nature and God. One cannot clap with one hand; one needs two hands to clap. Similarly Prakruti and Paramatma are essential for creation and without them, creation is impossible. Hence, it is sheer foolishness to consider Prakruti as a separate entity. The Samkhya system holds that Divinity runs as an undercurrent everywhere in Prakruti."

The yin and yang theory in Chinese philosophy also proposes this very same belief. The round symbol in fig. 45 illustrates this concept.

Fig. 45

Source: Wikipedia.

"The complementary dualistic concept in Taoism represents the reciprocal interaction throughout nature, related to a feedback loop, where opposing forces do not exchange in opposition but instead exchange reciprocally to promote stabilisation similar to homeostasis. An underlying principle in Taoism states that, within every independent entity, lies a part of its opposite. Within sickness lies health and vice versa. This is because all opposites are manifestation of the single Tao, and are therefore not independent from one another, but rather a variation of the same unifying force throughout all of nature."

Similar to the views expressed above, matter and antimatter are also exactly similar to one another except that they possess opposite charges. If we go according to this philosophy, then matter and antimatter have to coexist. In fact, without the presence of antimatter, even matter also should not have existed.

These views must not be taken lightly because unlike today in ancient times, many scholars and sages used to meditate days on end in quest of knowledge. It is a well-established fact that when knowledgeable persons continuously and single-mindedly concentrate their thoughts on a particular subject, they will ultimately be rewarded with answers and solutions to that they were in quest of. From time immemorial, humankind has always tried to find out the mysteries of life, our purpose of existence, and so on.

Hence, only the ancients could have succeeded in this philosophical quest, as philosophically they were far more advanced than we are or will ever be.

These opening lines from Einstein's 1920 lecture may help us in our search for antimatter.

"How does it come about that alongside of the idea of ponderable matter, which is derived by abstraction from everyday life, the physicists set the idea of the existence of another kind of matter, the ether?"

From our discussions on atoms, we know that other than the dense nucleus, they also contain a low-density substance that we named as the second matter. Logically speaking, this second matter is antimatter because this is what is supposedly missing from our universe. Just as the photons that constitute electromagnetic waves are a dilute form of matter contained in the nuclei of atoms the photons that make up gravither are also a dilute form of the second matter (antimatter) present in them. Matter represents energy, whereas antimatter represents the physical part of GM.

Although matter and antimatter might annihilate each other if they were to exist independently, nature has ingeniously ensured that this does not happen by making them into two entirely different entities, which however cannot exist independently.

This also vindicates the views and postulates of the illustrious personalities starting from Rene Descartes to Newton, Nicola Tesla, and others.

Isn't it strange and bewildering to know that, volume wise, our bodies and all the observable matter existing in the universe is mostly made up of a substance that everyone thinks is missing from our universe: the antimatter?

7
Misconception number 7

> *"The greatest enemy of knowledge is not ignorance it is the illusion of knowledge."*
>
> Stephen Hawking

Sun and stars contain mostly hydrogen and helium

The belief that the sun is comprised of mostly gases started from the time of Galileo.

According to the prevalent theory, this is based on the assumption that an independent gravitational force exists and our solar system formed from a huge ball of gas comprising of mostly hydrogen, helium, and a minute quantity of dust particles. This huge ball of gas spun so rapidly that it flattened out into a huge proto planetary disc from which the various planets evolved, whereas the sun itself evolved from the bulge at the center. If we go according to this theory, then either one of the two things mentioned below should have happened.

1) All the planets should have also been made up of only hydrogen and helium.
2) The sun must also contain all the elements present in the other planets in proportionately higher quantities.

As we are now aware that an independent attractive gravitational force emanating from celestial bodies does not exist and that it is the amount of mass and not the elementary composition of a celestial body which decides whether it transforms into a star or not, we are forced to accept the second option, which is that the sun must also contain, more or less, the same elements that are present in the earth and the other rocky planets in proportionately higher quantities.

There are a few other reasons, which indirectly support this presumption of ours.

1) As we are now familiar with how the gravitational force arises and functions around celestial bodies, the sun cannot be made up of only gases because their nature is such that the gas molecules, which possess high kinetic energy, will expand and diffuse into the vacuum of space instead of adhering to each other to form huge celestial bodies. Even today, if a huge quantity of highly compressed gas is released into outer space, it will instantly expand and disperse into space.

2) According to the current theory, elements having atomic number greater than iron are formed after a super nova has occurred. As supernovae are extremely rare events and take place billions of light years away from our solar system, the most puzzling factor is how the blown out matter reached here. The logistics involved in this vast journey through space are too incomprehensible. Even if we presume that, somehow it has managed to travel the vast distances and found its way here, why is it believed that it is not present in our sun, which accounts for more than 99% mass of our solar system. Our galaxy contains billions of stars and there are billions of such galaxies in the universe. Isn't it strange that our earth along with the other rocky planets including smaller bodies like asteroids contain almost all the elements present in the periodic table of element, whereas the trillions and trillions of stars that make up the universe are supposed to contain only hydrogen and helium?

3) The powerful magnetic field of the sun strongly implies that it must also contain metallic elements such as iron, which is the most abundant metal found on the rocky planets.

4) With the hydrogen-filled model of the sun, one cannot logically explain some of the various phenomena that occur on the sun. For example, the extreme temperature of the corona, coronal mass ejections and dark spots, and so on.

5) Experiments with chain nuclear fusion started in the 1950s have still not yielded the desired results although at that time scientists

were confident of achieving the task in 20 years or so. However, due to practical reasons, energy produced by nuclear fusion is still a long way off. From our earlier discussions, we know that the density of the sun's core is or is almost equal to the density of an atom's nucleus and it might measure thousands of kilometers in diameter making it behave like a massive uranium atom that is billions of times more radioactive. Although nature works in a highly repetitive manner, the results are truly amazing, astounding, and almost unbelievable. For example, the manner in which one element transforms into a completely different element with the addition of a single proton is almost magical. As it is the mass that triggers nuclear reaction both in an atom as well as the sun, logically speaking the same process taking place at the micro level, which is nuclear fission, must also occur at the macro level. The following analogy from the Upanishads aptly supports this line of thought.

"As is the microcosm, so is the macrocosm."

"As is the atom so is the universe."

May be it is because of these reasons that a few scientists are gradually noticing the weaknesses in the gas model of the sun and are coming up with alternate theories.

Alternate theories, experiments, and evidence

1) According to late Kristian Birkeland, a Norwegian scientist, who is also referred to as the first space scientist, mass is ejected from the sunspots during solar flares some of which end up orbiting the sun and slowly clump together to form planets. He considered the asteroid belt as masses halfway in the process from solar dust to planets. He conducted experiments using a Terrella, an iron sphere, placed inside a vacuum chamber along with electrodes through which he passed electricity. In the numerous experiments he conducted, he was successful in recreating the auroras that frequently occur near the arctic region and also simulated certain phenomena that occur on the sun's surface. He strongly believed that the sun definitely contained a large quantity of iron.

2) Michael Mozina, who strongly believes in Birkeland's theories, claims that he has studied numerous photographs taken by US satellites, which have been deployed with the specific purpose of investigating the sun. Some of these satellites are named as RHESSI, SOHO, ACE, TRACE, WIND, YOHKOH, and STEREO. He claims that these photos reveal the presence of a solid iron structure at a depth of about 4,800 km from the sun's photosphere.

According to his theory, nuclear reaction occurring in the sun's core releases protons and electrons, which rise to the surface, and when they come into contact with the iron layer, which acts as a conductor, tremendous amounts of electricity is generated and shoots up into the corona in the form of huge loops known as the coronal loops.

3) The following article was published under this headline.

Sun's X-rays reveal new twist (solar iron fusion)

According to new research from an international team at the Max Planck Institute of Nuclear Physics, Heidelberg, Germany, X-rays have revealed that although the sun is made up of mostly hydrogen and helium, there is a small but mighty iron core at its center. If this is true, according to the contemporary theories, the sun should be at end of its life.

4) Dr. Oliver Manuel, a professor teaching nuclear chemistry in the University of Missouri, USA, believes that the most abundant element present in the sun's interior is iron and the hydrogen present in its photosphere is a waste product. According to him, two founders of modern astronomy and astrophysics, Fred Hoyle and Arthur Eddington, were quoted as agreeing that sunlight is, "Chock-a block with lines of iron." He firmly believes that the hydrogen-filled model of the sun is obsolete. He also claims that many other eminent scientists suggest that the internal structure of stars mimics that of the atomic nucleus, which is exactly what this book claims.

5) Peter M. Bell an American scientist, late Carl Rouse an African-American scientist, late professor Paul Kazuo Kuroda an Japanese American scientist, late Dr. Charles Bruce an electrical engineer from England, and a host of others also believe that the sun contains a huge quantity of iron. They have also written many articles and books on the subject.

CHAPTER THREE

1
Forces that hold together bodies right from atoms to black holes

> *"As is the human body so is the cosmic body,*
> *As is the human mind so is the cosmic mind,*
> *As is the microcosm so is the macrocosm,*
> *As is the atom so is the universe."*
>
> Upanishad

Forces that hold together celestial bodies

The force that holds an atom together, which is known as the strong nuclear force, is the force arising out of the intense mutual repulsion between matter present in its nucleus and the SM present in the empty space surrounding the nucleus.

As an object comprising atoms grows in size, the intermolecular force initially holds them together. Later on, as the mass of the objects gradually increase, two new forces come into play. The force arising from the heaviness of matter (weight) always acts toward the center of mass of the body. The repulsion between the gravither membrane and the object gives rise to the compressive force, which acts on celestial bodies from outside. Initially, the intensity of the second force is just sufficient to prevent the gases from a heavenly body from escaping into space, and its contribution to the weight of objects on the surface of heavenly bodies is minimal.

However, as the temperature of a heavenly body increases due to increase in its mass, the intermolecular forces become progressively weaker. At this stage, a celestial body remains intact because of the tremendous downward force exerted by the increased weight of matter combined with the massive compressive force exerted by the gravither membrane. The ratio of contribution of the second force toward the weight of matter increases in proportion to the increase in mass of a celestial body until the body transforms into a gaseous state.

Combinations of the three forces namely the intermolecular force, force arising from the heaviness of matter, and the compressive force exerted by the gravither membrane (gravity) hold together all heavenly bodies right from a small asteroid to a black hole. The compressive force arising from the intense repulsion between matter contained in the nuclei and the second matter surrounding it holds the nuclei of atoms together.

An atom and a black hole are exactly similar to each other except for their size (i.e., mass).

2
Theory of everything

> *"Of course it would be a great advance if we could succeed in comprehending the gravitational field and the electromagnetic field together as one unified conformation. Then for the first time the epoch of theoretical physics founded by Faraday and Maxwell would reach a satisfactory conclusion. The contrast between ether and matter would fade away, and, through the general theory of relativity, the whole of physics would become a complete system of thought, like geometry, kinematics, and the theory of gravitation."*
>
> Albert Einstein

Standard model in physics

Before we study the TOE, let us briefly touch upon the Standard Model in particle physics. It is a theory concerning the electromagnetic, weak and strong nuclear interactions as well as the subatomic particles that constitute matter. According to the theory, the building blocks of matter are known as fermions. Protons, neutrons, and electrons fall under this category, and each of these are further constituted by smaller elementary particles like quarks, leptons, and so on. Particles such as the photons, gluons, and so on are classified in another group called the bosons. These topics come under quantum physics.

According to the Standard Model in physics, there are four fundamental forces of nature, which are as follows:

1) The electromagnetic force
2) The strong nuclear force
3) The weak nuclear force
4) Gravity

Out of the four forces, only the electromagnetic force and to a certain extent the weak nuclear forces are clearly defined. Although the theory is said to be very successful, it is dogged by the following shortcomings:

1) Currently, it does not include gravity.
2) It does not specify whether the strong nuclear force is an electromagnetic force or a mechanical one. Neither does it offer a scientific explanation for the extremely short range of this force.
3) Fails to explain how the addition or subtraction of a single proton can cause a new element to be formed.
4) Completely ignores the a priori presence of the low-density matter in all objects. Neither does it explain how the miniscule and super dense protons, neutrons, and electrons give rise to comparatively massive low-density visible, solid, liquid, and gaseous matter.
5) It fails to explain the irreversibility of the equation $e = mc^2$.
6) It fails to explain matter–antimatter asymmetry.
7) According to it, the mass of neutrinos should be zero. However, experiments have proved that they possess mass.

Due to these shortcomings, integrating all the four forces to illustrate a common origin or commonality is still a long way off.

Newly hypothesized TOE

In sharp contrast to the Standard Model in physics, the new theory propounded in this book not only integrates the strong nuclear force, the weak nuclear force, and gravity into a single force but also illustrates how all forces behave and function in the same manner. It also does away with the matter–antimatter asymmetry problem. It clearly illustrates how ordinary matter is constituted by two different entities; one representing the energetic part of matter and the other the physical part. The nucleus, which is a condensate of electromagnetic energy, represents matter and the second matter surrounding it, which is a condensate of the gravither present in outer space represents antimatter.

Likewise, according to the new hypothesis, there are basically only two fundamental forces, which are as follows:

1) The electromagnetic force, which arises from the nuclei of atoms.
2) The nonelectromagnetic force arising from the mutual repulsion between matter contained in the nuclei of atoms and the second matter (SM) surrounding them and the mutual repulsion between celestial bodies and the gravither membrane. The strong nuclear force, the weak nuclear force, and gravity are manifestations of this nonelectromagnetic force, which means that all these three forces are mechanical in nature.

Because of this difference between the two forces, the electromagnetic force cannot be integrated with the other three fundamental forces.

The intra and intermolecular forces can be treated as the third fundamental force as they are mechanical forces arising from the stickiness of the second matter and its tendency to liquefy when heated and harden when cooled.

If we observe carefully, we will notice that there is nothing very new or extraordinary in this new hypothesis because in reality, it is a holistic combination of Rutherford's atomic theory and the ether theories propounded by Rene Descartes, Sir Isaac Newton, James Maxwell, Nikola Tesla, and a host of other equally illustrious personalities. Just as Rutherford's theory is supported by strong experimental proof, the ether theories are supported by the following phenomena; the a priori presence of the low-density matter in all bodies, the heating up and bouncing of spacecrafts at an altitude of 120 km from the earth's surface, and so on. Moreover, if the experiments mentioned in this book are carried out, more supporting evidence is bound to emerge.

The following is a quote by Mr. Nainan K. Varghese, who has authored a book titled, *A Hypothesis on Matter*.

"It is a long-time dream of scientists to have a unified theory, which is applicable under all conditions and to all phenomena that are related to matter. This desire clearly indicates that all is not well with the current theories and hence the search for a new 'holy grail' of theory. These searches, in some way or other way, are based on present-day theories or they are continuations of present-day theories. Searches based on 'not very correct theories' cannot, yield a better theory, they can only compound errors. Hence scientists trying to find a better theory, need to go outside the thinking of orthodox physics. This they are afraid to do. Anybody, who dares to move outside orthodox physics in search for such a theory is immediately branded as heretics and are isolated from the mainstream. Their work is not given any consideration. This tendency on the part of the establishment is the main reason for the delay in any progress toward a unified theory."

Conclusion

As we have observed, our universe is very much physical in nature and nature itself works on well-established natural laws. Celestial bodies are made up of the incredibly tiny particles called atoms and atoms themselves are made of even smaller subatomic particles. Even energy in the form of light and all the other electromagnetic waves consists of tiny particles called photons. Irrespective of the strange and unpredictable behavior of matter at the quantum level and at the speed of light, everything in the real physical world that we live is sensible and predictable. There is neither magic nor any other sort of mystical phenomena in nature. Behind every phenomenon, whether physical or otherwise, there is always a scientific cause or explanation.

Yes, it is indeed true that nature never deceives us; it is we who deceive ourselves and as Newton has rightly pointed out truth is ever to be found in simplicity and not in the complexity and multitude of things. Why will nature resort to such complicated methods, a mystical force endowing mass and an equally mystical and magical force called gravity endowing weight to matter when the same can be easily achieved by directly endowing matter with these properties? That is why the inertia of an object is entirely dependent on its weight. The heavier an object is the greater the inertia and vice versa.

Purely, theoretical physics is as important as mathematics and equations. Imagine a person who is a brilliant theoretician and another who is a brilliant mathematician. Individually, they may not be able to achieve much, but just think what would happen if both of them were to work together complementing one another. The result would be astounding. Moreover, theoretical physics forces us to think and act in a logical and rational manner as all phenomena must be explained in a logical and scientific manner. Ignoring Newton's gravitational ether theory along with Einstein's 1920 Leiden lecture and reducing the mechanical views of nature to purely that of electromagnetism are monumental blunders that have deeply hindered the progress of fundamental physics. Alternate theories should also been given a fair hearing instead of straightaway banning them because if something is untrue, it will eventually be noticed and discarded. On the other hand, truth, however much it is suppressed, will eventually raise its sublime head and reveal itself to one and all much to the consternation of those involved in suppressing it.

3

Inferences

The new hypotheses posited in this book lead to the following inferences

1) Energy in any form can never be converted into the ordinary low-density matter, present in all objects, because they are two entirely different entities.

2) An independent ever-attractive gravitational force will never be discovered because such a force does not exist anywhere in the universe. Hence, both Einstein and Newton's later postulate, which doubted the existence of such a force are indeed true.

3) Gravity waves posited by Einstein, if mechanical in nature, will never be detected as the elastic medium (gravither) pervading space will dampen and absorb all mechanical vibrations due to which, unlike electromagnetic waves, they cannot travel long distances in space.

4) If a sufficient quantity of the gravither present in space is collected and compressed, it will transform into a transparent rubber-like substance, possessing antigravity properties.

5) Antimatter in physical form will never be discovered because it already exists in the form of the low-density GM found in all objects and as gravither in outer space.

6) There is a sixth inference also, which is to do with the happenings inside the sun's core, which I am sure that you can figure it out on your own.

Appeal

I once again appeal to all my readers to apply the principle of Occam's razor and decide for yourselves if the new theories and hypotheses posited in this book, which are in fact a reiteration of the opinions and theories of the illustrious, eminent, and most accomplished scientists, physicists, and scholars of yesteryears, holds any water and pass your judgment.

Real or fiction

Recapitulating the list of already existing evidence and the proposed experiments

Existing evidence

1) Matter contained in all objects is anything but empty. Even a blind man will vouch for this.
2) Bouncing and heating up of spacecrafts at exactly 120 km from the earth's surface.
3) Spherical formation of liquid bodies in space.
4) Weightlessness even inside sounding rockets, which do not go into regular orbits, while they are above 100-km altitude from the earth's surface.
5) Comparing earth's atmospheric layers with that of the sun.

Practical experiments

1) Experiment using a pneumatic cylinder to find out if outer space is filled with an elastic medium.
2) Experiment to find out if weightlessness experienced inside spacecrafts is due to absence of the space medium gravither inside it.
3) Experiment to check if an object in space is still in possession of its weight.
4) Experiment to prove that gravitational fields arise around objects only when they are in outer space.
5) Experiment to detect if a downward force is acting on the earth's surface from space.

6) Compressing hydrogen cations or alpha particles to see if they transform into a liquid.

7) Using Einstein's thought experiment to find out if an attractive gravitational force really exists.

8) Conduct a Michelson–Morley type experiment in space.

9) Experiment to check the antigravity properties of the gravither.

10) Experiment to compress gravither and find out if it transforms into a physical entity.

11) Firing a rocket vertically upwards to find out if thrust required to overcome its mass decreases with increase in altitude or stays constant.

Answers, wherever possible, to the various queries raised at the beginning of the book

Questions on atoms

1) Nature has created everything existing in this Universe with a specific purpose and task. Hence, what is the purpose of the empty space present in atoms?

   Answer: The vast empty space is not really empty but contains a low density substance which mutually repels with the nucleus due to which the nucleus, that represents electromagnetic energy, is compressed to a highly dense state.

2) Is the empty space in an atom the same as that existing in outer space?

   Answer: The empty space in an atom is not really empty but contains a condensed form of the ethereal and elastic substance (gravither) that is present in the continuum of outer space.

3) When atoms of lighter elements, which are in a free state, fuse together to form atoms of heavier elements, their volume does not increase proportionately. What happens to the empty space associated with the atoms of the lighter elements after they fuse together?

   Answer: The elastic medium present in the empty space of atoms fuse together due to which the density of the heavier element increases. That is why the volume of the newly formed elements does not increase proportionately.

4) What happens to the empty space associated with a massive star that transforms into a neutron star or a stellar black hole measuring only 10 to 15 km in diameter after a supernova has occurred?

   Answer: The empty space associated with the massive star, which is full of condensed gravither, will become a part of outer space after the supernova has occurred.

5) If an atom consists of mostly empty space, what prevents objects from passing through each other?

   Answer: Objects are prevented from passing through each other because the supposedly empty space is not really empty but full of a low-density substance.

6) From when and on what basis exactly is it believed that ordinary matter is a condensate of electromagnetic energy only and if this assumption is true then, why can't electromagnetic energy be converted into ordinary matter?

   Answer: From the early 19th century matter was believed to be a condensate of only electromagnetic energy as the various experiments failed to detect the presence of ether. Electromagnetic energy cannot be converted into ordinary matter because they are two entirely different entities.

7) If atoms consist of only protons, neutrons, electrons and 99.99% of empty space, why do objects appear as being full and what is the low-density substance present in all objects? Is the FBBTS (flesh, blood, bones, tissue and skin) present in our bodies just an illusion created by the whirring electrons or electron clouds?

   Answer: Objects appear as full because they are really filled with a low-density substance and not because of any illusion.

8) Each addition of a proton to an element results in the formation of the next element with a higher atomic number. What is the internal mechanism or process by which this transformation takes place?

   Answer: When protons are added to an element, the increased energy radiated by the nucleus is absorbed by the second matter, which causes it to transform into heavier elements.

9) When an object, like wood, burns what exactly is burning? Is it the protons, neutrons or electrons?

   Answer: It is the second matter surrounding the nucleus, which burns and not the protons, neutrons or electrons.

10) Since, as per current theories, electrons, which are negatively charged and form the outer shells of atoms can exist either as point particles spinning rapidly around the nuclei or as electron clouds how do atoms manage to stick to each other?

    Answer: Atoms and molecules stick together because of the adhesive nature of the second matter and not because of electromagnetic forces.

11) Why is the range of the strong nuclear force, which is billions of times stronger than gravity, so unimaginably and ridiculously short? It is effective only within the nucleus of an atom.

Answer: The range of the strong nuclear is so short because it is not a force emerging from within the nucleus but a compressive force that acts on it from outside because of the mutual repulsion between the nucleus and the second matter surrounding it.

12) What type of a force is the strong nuclear force electromagnetic, electrostatic or mechanical?

Answer: The strong nuclear force is a mechanical force.

Questions on gravity and outer space

1) Why is matter, by itself, considered weightless?

    Answer: Matter by itself is considered to be weightless because huge celestial bodies appear to float in space.

2) How exactly does mass give rise to the force of gravity or the gravitational field?

    Answer: The mass contained in a celestial body displaces the gravither present in outer space. This causes the density of the gravither surrounding it to increase in a gradient manner only till a certain distance from the celestial bodies surface. This forms the celestial body's gravitational field.

3) Under which category is the gravitational force classified, mechanical or electromagnetic?

    Answer: The gravitational force should be classified as a mechanical force.

4) Mass is defined as the amount of matter contained in an object. It is also defined as the resistance an object offers to being accelerated when it is subject to a force. Even in a gravity free environment, we have to apply a force to overcome inertia and move the object. What characteristic or property of matter gives rise to this resistance or inertia?

    Answer: It is the heaviness of the matter contained in a body that give rise to inertia. The heavier a body is the greater its inertia and vice versa.

5) Think of a rocket whose mass is say 100 tons and weight 980 KN. As its distance from the earth's surface increases its weight, as per current theories, will gradually decrease as the strength of gravity is also decreasing and at a particular altitude, say 1.5 million km, its weight will be almost zero although mass remains unchanged at 100 tons. Hence, theoretically, a thrust of only 3 to 4 N or even less is sufficient to propel the rocket upwards. However, since the rocket is still in possession of its mass is it not necessary to apply an appropriate force to overcome its inertia in the upward direction also or is inertia applicable only in the horizontal direction?

   Answer: Since inertia is not dependent on directions and the mass of the rocket remains constant at all altitudes the thrust required to propel it upwards will always remain proportionate to its mass and will not reduce with altitude increase even though weight may appear to reduce.

6) Tremendous amount of energy is required to accelerate any object. Since gravity is supposed to be a force, which causes freely falling objects to accelerate what is the source of this energy and if such a force really exists what is reason behind its non-detection?

   Answer: Since an independent attractive force does not exist the question of energy does not arise. Objects fall down freely because of the heaviness of the matter contained in them and accelerate because of the potential energy stored in them.

7) How can the range of gravity be infinity if it is the weakest of the four fundamental forces?

   Answer: Once again, the range of gravity becomes irrelevant as such a force does not exist. However, the gravitational field has a definite range and this is directly proportional to the mass of a celestial body and its distance from the sun.

8) Why are most celestial bodies spherical in shape and irrespective of their mass why do liquid bodies also attain a spherical shape in outer space?

   Answer: Most celestial bodies are spherical in shape because they must have been in a molten state at the time of their formation. Liquid bodies take on a spherical shape in space because in outer space, there are no upwards and downwards directions due to which they behave as independent celestial bodies. The heaviness of matter from all directions will act towards the centre of mass of that body due to which it attains a spherical shape as the molecules in the liquid body can slide against each other. Moreover, the gravither present in space also exerts a compressive force on it from outside.

9) Why liquid bodies do not assume a spherical shape below 100 km altitude from the earth's surface?

Answer: Below hundred kilometers altitude bodies do not behave like independent bodies due to which the heaviness of matter contained in them will not act towards their own centre of mass but will instead act towards the centre of mass of the earth. Since there is no gravither below 100 km altitude the compressive force also is absent.

10) Why does weightlessness or microgravity occur only above 100 km altitude from the earth's surface?

Answer: All bodies entering space don't experience micro gravity. That is why sounding rockets and other spacecrafts fall down towards the earth when their rocket engines are shut off or when their orbital speed decrease beyond required limits.

11) 99.99997% of atmospheric air exists below 100 km altitude from the earths' surface and the remaining 0.00003% spreads out in thousands of kilometers above this altitude implying that outer space is just an empty vacuum. If this is true, why do spacecrafts encounter severe orbit decay, which is highest just above 100 km altitude and decreases with increase in altitude?

Answer: Orbit decay has two causes. First, the density of the gravither is maximum at 100 km altitude above which it decreases as the altitude increases. Hence as spacecrafts move to higher altitudes the friction decreases. Secondly, the ups and downs of the earth's surface is superimposed onto the gravither present is space. Even this is smoothened out as the altitude increases until at a certain altitude the density of the gravither is uniform.

12) During re-entry, why do spacecrafts heat up exactly at 120 km altitude from the earth's surface where the density of air is almost negligible?

Answer: Spacecrafts reentering at high speeds heat up when they enter the gravither membrane because the higher density of the gravither membrane gives rise to severe friction.

13) Similarly, during re-entry why do spacecrafts, like the shuttle weighing hundreds of tons, bounce back into space at exactly 120 km altitude from the earth's surface if their angle of attack (AOA) is less than 40 degrees?

    Answer: Since the spacecraft is reentering at 27000 km with an approach angle of about one degree it bounces back into space because of surface tension. The gravither membrane beginning at 120 km behaves like an elastic membrane similar to how a placid lakes surface behaves when a stone is thrown at a very shallow angle.

14) Why do objects and astronauts inside a sounding rocket, which does not go into a regular orbit, experience weightlessness in space even while ascending?

    Answer: Every object that is inside a spacecraft that is in space will experience weightlessness, as long as all the rocket engines are shut off, because the gravither present is space is not present inside the spacecraft. The speed and direction of the spacecraft have no bearing on the weightlessness experienced by the objects.

15) The strength of gravity on the moon's surface is one-sixth the value of earth's gravity. Since numerous manned and unmanned missions have landed on the moon, did the astronauts conduct any direct weighing experiment to verify this postulate? If yes what was the result and if not, why did they not conduct it?

    Answer: As per the new hypothesis, since a gravitational force does not increase, weight of an object of unit mass will increase only if its density increases. Hence the weight of earthly objects on the moon will be the same as on earth or very slightly less.

16) When a student questions how gravity can bend light if the rest mass of photons is zero (gravity acts only on bodies or particles that possess mass) he gets the following reply. Since energy and mass are equivalent, gravity is able to bend light. However, since this explanation violates the laws of quantum mechanics, according to which gravitational

interaction at the quantum level is nonexistent, how does one explain this dichotomy? The photons are so unimaginably small that the total number which travel from the sun and strike us every second are actually spread across a length of 300,000 km, which is light's speed. Moreover, if this assumption that the photons are zero mass particles is true, where is the suns' mass disappearing and how are they able to gain momentum to travel at the speed of light?

Answer: As hypothesized by Newton the gradient density ether surrounding huge celestial bodies like the sun causes light to bend due to refraction.

17) Is space-time a physical entity and if yes how does mass affect its curvature? If it is not a physical entity then what exactly is it?

Answer: Space-time is not a physical entity. Einstein himself has made this clear, in his 1920 lecture, by expressing repeatedly that space is endowed with physical qualities because it is filled with ether, which is a physical entity similar to gases.

18) Why are most contemporary scientists dead scared of ether theories? After all, if it is there it is there and if not no. From when and why are these theories banned?

Answer: People fear to talk about ether theories as this might jeopardize their carriers. The ether theories have unofficially been banned from the late nineteen thirties.

Questions pertaining to our solar system and general questions

1) Why are huge celestial bodies always gaseous in nature and smaller heavenly bodies like earth always rocky?

    Answer: As the mass of celestial bodies increase, their dense cores begin to behave like the nuclei of heavy elements like uranium 235, which undergo nuclear fission. The lighter gasses released during this process rise up to the surface giving rise to the impression that the entire celestial body is comprised of only gases.

2) Why do scientists believe that antimatter is missing from our Universe?

    Answer: Because they think that ordinary matter is made up of a single entity which is a condensation of electromagnetic energy only.

3) Earth, a ridiculously tiny celestial body, contains all the elements present in the periodic table of elements, whereas the trillions of stars present in the billions of galaxies are supposed to contain only hydrogen and helium along with a few traces of lighter elements. What could be the reason for this strange anomaly?

Answer: Since the sun and all the planets have formed from the same cloud of gases or whatever logically speaking the sun and all stars must also contain the same substances as found on earth and the other planets.

4) Distant planets like Jupiter, Saturn, Uranus, and Neptune are releasing more thermal energy than they receive from the sun. What is the source of this excess energy?

Answer: Excess energy is being released because nuclear fission is occurring in their cores.

5) As per the theory of the birth and evolution of our universe, a number of non-metallic stars (in astronomy other than hydrogen and helium all other elements are metallic) should have been present in the Universe. Hence, why have astrophysicists been unable to discover a single non-metallic star even until today?

Answer: Non-metallic stars have not been discovered because all stars contain more or less the same substances present on the earth but in different proportions. However, this is hidden by the thick layer of gases surrounding them.

6) Does time really exist or is it just a notion?

Answer: Time is not a physical entity but just a notion.

7) Light by itself is invisible. However, when it strikes an object, not only does it become visible, it also generates heat. What exactly happens when it strikes an object?

Answer: Light consists of particles called photons travelling at 300000 km/sec. Irrespective of whether they are in possession of mass or not when they strike an objects surface, they heat up tremendously due to friction because of which heat and light is generated. This is exactly similar to how a bullet behaves when it strikes a hard surface at very high speeds.

5
Speed of light

One of the most famous postulates of the special theory of relativity is that the speed of light is constant irrespective of whether its source is in motion or at a standstill.

Since this postulate is contrary to what we experience and observe in the real physical world, let us try to find out how this can be possible.

Let us first discuss about relative motion by visualizing a boy, with a ball in his hand, standing on a train, which is moving at a speed of 30 km/hr. If he throws the ball with a speed of 40 km/hr in the direction of the train's movement to an observer standing motionless on the ground, it will appear to move at 70 km/hr, which is the sum of the two speeds and which is the true speed of the ball with respect to the earth's surface. If he now throws the ball in the opposite direction, that is, against the direction of the train's motion with the earlier speed, to the observer on the ground, it will now appear to move at only 10 km/hr, which is equal to the speed of the ball minus the train's speed.

Let us now visualize that instead of the ball the boy is holding a light emitting torch and points it first in the direction of the train's movement and then in the opposite direction. According to the constant speed of light theory, the speed of light unlike in the case of the ball will appear unchanged to the external observer.

To get a better idea as to what is happening, let us visualize a spaceship that is moving at 200,000 km/sec, just imagine for the experiments sake that such a vehicle exists and has a powerful torch mounted on it. If the torch is pointed in the direction of the spaceship's motion and switched on to an observer on the ground, the speed of light will still appear to be 300,000 km/sec, whereas logically, it should have been 500,000 km/sec. Even when the torch is pointed in the opposite direction, the speed of light will still appear to be 300,000 km/sec to the observer instead of only 100,000 km/sec.

There is only one way in which this is possible and that is the speed of light emerging from the torch has to slow down to 100,000 km/sec in the first instance and speed up to 500,000 km/hr in the second case. There is absolutely no other possibility by which this phenomenon can occur.

So far, we had only one external observer who was moreover stationary. Now see how it becomes even more bizarre when we have a number of observers randomly moving in opposite directions at different speeds. Now irrespective of the speed of the aircraft and the speed of the various observers, the speed of light should simultaneously appear to all of them as 300,000 km/sec.

To regulate something (e.g., the maximum speed of a car or the temperature of a furnace), a sensing mechanism, a feedback loop, and a device, which will increase or decrease the energy required is absolutely necessary. The maximum speed limit of automobiles can be controlled by the use of a governor. In the case of furnaces, the temperature is controlled with the use of thermocouples coupled with other devices. However, as light is neither an intelligent entity nor is there any feedback loop, how is light able to control and regulate its own speed?

Because this is a matter of fundamental importance, it would be very interesting and also useful to discover how light is able to behave in such a mysterious manner. I hope that someone in the near future throws some light on this matter.

Epilogue

> *"Everything is determined, the beginning as well as the end, by forces over which we have no control. It is determined for the insect, as well as for the star. Human beings, vegetables or cosmic dust, we all dance to a mysterious tune, intoned in the distance by an invisible piper."*
>
> Albert Einstein

It is now very clear that the whole universe is a product of the energies of matter and antimatter (gravither) and nothing else.

Dualism is the bedrock on which the universe has formed, evolved, and is functioning. Every particle has an opposite particle, and likewise, every force has an opposite force. Without darkness, we will not know what light is, and without evil, we will not know the meaning of good. Even in our real lives, this dualism is clearly evident. Procreation cannot be achieved by a single entity. Instead, both the opposites are required. Likewise, both good and bad qualities are within us, and hence, no person is ever completely bad or completely good. The universe that appears as made of a single entity is in reality a harmonious blend of two different types of matter or energies, which although opposite are highly interdependent. Right from an atom to black hole, antimatter always surrounds matter, and hence we can say that antimatter always dominates over matter everywhere in the universe. Without antimatter, matter would exist only as radiation in space.

On the other hand, antimatter is omnipresent throughout the universe. It is this which binds not only the nucleus of an atom but also individual galaxies together. It is as though the essence of God is present in this exotic, ethereal, and divine material.

If we now recall the yin and yang theory, we can say that yin is matter and represents the female component, whereas yang is antimatter and represents the male component. Matter always represents energy, whereas antimatter acts as the housing for that energy. This may also be the reason why, in our civilization including that of animals and birds, males always dominate females.

All forces including gravity and all objects including humans are products of these two oppositely charged energies. The whole universe, which has formed after billions of years of evolution, can easily disintegrate and vanish into empty space (as radiation) in a fraction of a second, if the mutual repulsion between matter and second matter were to cease even for a few seconds. Everything will be back to the beginning of the universe.

All heavenly bodies are like natural engineless spacecrafts, surrounded by the gravither membrane and the gravitational fields, which serve as the outer shell of these natural spacecrafts and protect them from the harmful effects of outer space. These natural spacecrafts, of which our earth is also one, are like perpetual motion machines silently and diligently moving in their designated paths for billions of years.

According to Charles Darwin, life on earth evolved by itself and God had no hand in it. He is probably correct in saying so. However, from where and how the earth, on which conditions are so favorable for the evolution of life, come into existence is a question that cannot be answered by Charles Darwin's theory of evolution.

Right from a black hole to our present state, there has been not just evolution, but a meticulously pre-planned evolution and human beings are at the pinnacle of this pre-planned evolution. It is impossible for any human being to suggest or even imagine that the whole universe and life on earth evolved all by itself without the involvement of a divine force.

This also proves that there exist two supreme or divine energies in the universe, which have not only done this meticulous pre-planning from a micro to a macro level, but have also created and executed it to unbelievable perfection. These supreme and divine cosmic energies are what we humans refer to as the Omnipresent, Omniscient, and Omnipotent God. However, referring to god as a male entity is nothing but male chauvinism. There must be two entities: one representing the male and the other female.

Now let us discuss a little about ourselves. From time immemorial, scholars, philosophers, and even ordinary people have wondered about the purpose of our existence on this earth. Until now, we have not been able to concisely define or specify what consciousness really is. The answer probably lies in this simple but highly philosophical and wise statement from the Upanishads *"as is the human mind so is the cosmic mind."* Boredom and loneliness are something that terrifies everyone, and hence, we are always on the lookout for variety whether it is concerned with food or any other social activity. Scientists are trying to mimic God by trying to create new life and have presently achieved cloning. Similarly, they have manufactured robots with artificial intelligence. Strangely, we humans come into this world not by choice but by chance. Because nothing is in our control right from birth until death, we are in a way similar to the artificial robots with one significant difference. We are highly sophisticated, self-perpetuating, self-sustaining, and intelligent biological robots with the ability to express and feel all sorts of emotions. Probably, just like us, the mysterious entities that we refer to as God also feel bored and lonely in the vast cosmos and have created us to get rid of these feelings. They are not sitting somewhere in the skies or heavens and watching us but are a part of us in the form of the mysterious consciousness. Through us, they feel the various emotions like joy and sorrow, ecstasy and pain, and so on. The world is a big live drama theatre in which we are the actors. Some have good roles and others bad. There is comedy, tragedy, romance, suspense, murder, robbery, and so on just as in the artificial drama theatres. Just as the acting is not real and actors go back to their normal life after acting, our lives are also not permanent and real but temporary. Our bodies, which are made up of lifeless materials found in the earth, have value only until we are alive. After death, our bodies rot and become a part of the earth, and hence, the saying in the Bible: *"dust to dust and ashes to ashes."* However, our consciousness, which is connected to our bodies at a spiritual level, never dies but reverts to its origin the mind of God.

Glossary

Atom: The smallest part of an element consisting of a dense positively charged nucleus with negatively charged electrons orbiting around it.

Antimatter: Replica of matter, but possessing opposite electrical charge.

Black hole: A huge celestial body whose mass is billions of times more than sun and is located in the center of galaxies.

Electron: Negatively charged particles present in atoms whose mass is almost 2,000 times lesser than that of a proton.

Gravither: A new name proposed for the elastic medium pervading space.

Gravither membrane: Protective elastic layer found around celestial bodies formed by the displacement of gravither.

Geodesic: The shortest path between two points on a curved surface.

Gravity: Ever-attractive fundamental force believed to exist between two bodies and which also endows weight to matter.

Neutron: Neutrally charged particles present in the nucleus of atoms whose mass is fractionally more than that of a proton.

Proton: Positively charged particle present in an atom's nucleus, which is the basic building block of all matter.

Photon: Mass-less particles that constitute light.

Positron: Antiparticle of electron possessing positive electric charge.

Radioactivity: Breakdown of one type of atomic nucleus into another of lower atomic number accompanied by radiation.

References

1) Wikipedia and Wiki quotes.

2) Sandi Fer, E. How Euler Did It.

3) Galili, I., & Tseitlin, M. Excurse to the History of Weight Concept: From Aristotle to Newton and then Einstein.

4) Einstein, A. Lecture on the Topic of Ether and Relativity.

5) Ranzan, C. The History of the Ether Theory.

6) Collation of Scientific Theories of the Ether. Australia: Mountainman Graphics.

7) Gorbatsavich, F.F. The Ether and Universe.

8) Shank, J.B. The Newton Wars and the Beginning of French Enlightenment. (pg. 47).

9) Manuel, O.K. My Journey into the Core of the Sun.

10) Bruce, C. The Surface of the Sun.

11) Bell, P.M. Iron Core in the Sun.

www.ingramcontent.com/pod-product-compliance
Lightning Source LLC
Chambersburg PA
CBHW050205230526
45470CB00001B/244